T0297909

An Elementary Introduction to

QUEUEING
SYSTEMS

An Elementary Introduction to

QUEUEING SYSTEMS

Wah Chun Chan

University of Calgary, Canada

World Scientific

NEW JERSEY · LONDON · SINGAPORE · BEIJING · SHANGHAI · HONG KONG · TAIPEI · CHENNAI

Published by

World Scientific Publishing Co. Pte. Ltd.
5 Toh Tuck Link, Singapore 596224
USA office: 27 Warren Street, Suite 401-402, Hackensack, NJ 07601
UK office: 57 Shelton Street, Covent Garden, London WC2H 9HE

British Library Cataloguing-in-Publication Data
A catalogue record for this book is available from the British Library.

AN ELEMENTARY INTRODUCTION TO QUEUEING SYSTEMS

ISBN 978-981-4612-00-5

Printed in Singapore

*This book is dedicated to the memory of my uncle and aunt,
Mr. and Mrs. Lap Hoi Chan, who supported me during my youth,
and my professor, Dr. Donald A. George, who inspired me
in the study of the theory of probability.*

ACKNOWLEDGEMENTS

The author wishes to thank his wife, Yu-Chih, and his family members, Eileen and Al, Jean and Aaron, Vivian and Brian, and An-Wen for their encouragement and support during the preparation of the book. Also, a special thanks to Eileen for her skillful typing of the manuscript in her busy work schedule.

CONTENTS

PREFACE

Societal interactions often involve situations where people must wait for service. Examples include queues in shops, ticket offices, hospitals, et cetera. The study of these phenomena is known as queueing theory. Any system in which customer arrivals demand service from a limited number of servers can be called a queueing system. In all practical situations, the arrival times of customers are unpredictable. The main task of queueing theory is to establish the interdependence of the number of servers and the quality of service. The quality of service in different situations is measured by different indices, usually either the percentage of demands that are refused or the average waiting time for the beginning of the service. Obviously, a higher quality of service requires a greater number of servers. However, it is evident that an excessive number of servers will result in wasted resources. Thus, in practice, the problem is usually resolved by determining the minimal number of servers to achieve the desired quality of service.

In problems of queueing, it is always necessary to account for the influence of uncertainty on the course of the phenomenon under consideration. The rate and behavior of customer arrivals are not, as a rule, completely known. In addition, the service times vary randomly from one problem to another. All these chance elements constitute the main features in the processes to be studied. Thus, it is natural that the concepts and methods of the theory of probability should become a mathematical necessity for the study of queueing systems.

The purpose of the present work is to acquaint readers with the main concepts, methods, and approaches that facilitate the application of probability to problems of queueing systems. A book of this elementary treatment should be most useful for students and practising engineers who wish to understand the fundamental characteristics of the most important queueing systems. My objective has been to employ the birth and death process as a basic mathematical model for the investigation of queueing systems whenever applicable. Other concepts and methods are also introduced using probability arguments, such as the method of imbedded Markov chains. Whenever possible, I have tried to use capital letters to denote random variables and small letters their values.

This book is divided into five chapters. The fundamental nature of the Poisson input process and the birth and death process are discussed in Chapter 1. Queueing systems with losses (the Erlang loss system) are investigated in Chapter 2. The presentation is simplified by employing the birth and death process as a model for the system. In Chapter 3, the investigation of queueing systems allowing waiting is carried out by using the results of the birth and death process as a model. Further, the distribution function of waiting time is determined. Chapter 4 deals with the study of the Engset loss and delay systems. The last chapter studies some non-Markovian single server queueing systems.

CHAPTER 1

MODELING OF QUEUEING SYSTEMS

1.1 Mathematical Modeling

In the analysis of a physical system, the first step is to derive a mathematical model for the system. However, it is important to note that mathematical models, in general, may assume many different forms. Depending on the particular system, one mathematical model may be more suitable than others.

To begin our discussion, let us consider a physical phenomenon. Customers arrive and request a certain kind of service. If a server is available, the arriving customer will be served for a certain length of time, after which the customer will depart and the server will become idle and be available for other customers. If the arriving customer finds no idle server, he will wait in a line (queue). This phenomenon may be depicted by the diagram in Fig. 1-1.

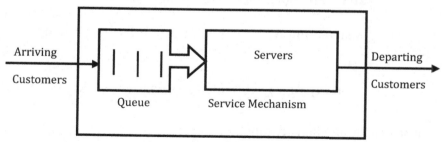

Arriving Customers

Servers

Departing Customers

Queue   Service Mechanism

Fig. 1-1. A queueing system.

Under certain idealized assumptions, many queueing systems may be characterized by random processes, such as the input of arriving customers and the service times of servers. The statistical behavior of a queueing system may be obtained by relating these random processes. The mathematical description of the characteristics of the queueing system is called a queueing model. Deriving a reasonably accurate mathematical model for a queueing system is the most important part of the entire analysis. The study of queueing systems is concerned with the analysis of the mathematical models representing the real physical queueing phenomena. The development of the theory of queueing has its origins in the study of congestion in telephone systems [6].

The principal characteristics of queueing systems are: (a) the input process, (b) the service mechanism, and (c) the queue discipline. These characteristics were proposed by D. G. Kendall in 1951 [1] and now are widely used to describe queueing models. In Kendall's short hand notation, we use A/B/C to denote the arrival (input) process, the service time and the number of servers respectively. Also, a modified form of Kendall's notation A/B/C/D/E has been used, where D specifies the maximum number of customers who may be present in the system at any one time (including those being served), and E specifies the queue discipline.

Some examples of Kendall's notion are:

M/M/m   represents a queueing system with Poisson input, exponential service times, and m servers;

M/G/1   denotes a queueing system with Poisson input, general (arbitrary) service times and one server;

$GI/E_k/m$ denotes general, independently distributed interarrival times, Erlangian service times and m servers.

The Erlang distribution of order k with rate μ has the distribution function

$$E_k(x) = 1 - \sum_{j=0}^{k-1} \left(\frac{(\mu x)^j}{j!} \right) e^{-\mu x}, \text{ where } x \geq 0, k \geq 1$$

whose probability density function is

$$e_k(x) = \left(\frac{\mu^k x^{k-1}}{(k-1)!} \right) e^{-\mu x}, \text{ where } x \geq 0, k \geq 0$$

In the study of queueing systems, it is always assumed, for mathematical convenience, that the waiting capacity of the queue is infinite. In addition to Kendall's terminology, the queue discipline is usually specified separately. The following notations are commonly used:

FIFO   represents first into the queue and first out of it into service, or first-come, first-served discipline, or service in order of arrival;

SIRO   means service in random order, or customers are selected randomly from the queue to obtain service;

LIFO   denotes last-come, first-served discipline.

Of all the queue disciplines the FIFO discipline is the most natural one and is the fairest from the point of view of customers, but it may not be the best from the point of view of the servers.

1.2 The Poisson Input Process

The input process of a queueing system consists of the flow of incoming customers in an orderly manner. Our objective will be to find a mathematical representation of the input process. Note that the incoming customers arrive in the queueing system randomly. The time epochs at which individual customers are seen at the input T_i of the system are called arrival epochs. The intervals between any two consecutive arrival epochs are called interarrival times. These quantities are depicted in Fig. 1-2.

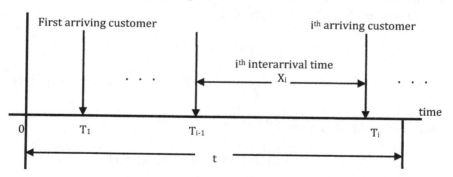

Fig. 1-2. The arrival epochs and the interarrival times.

Since customers arrive in a random or unpredictable manner, the arrival epochs T_i, i = 1, 2, ... , are clearly random variables, as are the interarrival times $X_i = T_i - T_{i-1}$. The specification of the probabilistic feature of the input process may be made in terms of the distribution of the length of X_i or the distribution of the number of arrivals in a fixed period of time of length t, regarded as a random variable N(t).

Now let us consider the random variable N(t) in the time period t and then divide t into n small and equal subintervals

$$\Delta t = \frac{t}{n} \tag{1.1}$$

If Δt is sufficiently small such that there is either no or only one arrival in it, then the number of arrivals in t can be any integral value from 0 to n. The important quantity we need to calculate is the probability of exactly k arrivals in the period of length t. Let this probability be denoted by

$$P_k(t) = P\{N(t) = k\}, \quad k = 0, 1, 2, ... \tag{1.2}$$

Suppose that the total number of arrivals in t is n_A. Then the average arrival rate is defined by

$$\lambda = \frac{n_A}{t} \tag{1.3}$$

Using the statistical (relative frequency) definition of probability, an arrival is found in Δt with probability

$$P_1(\Delta t) = P\{N(\Delta t) = 1\} = \frac{n_A}{n}$$

$$= \frac{n_A}{t} \times \frac{t}{n} = \lambda \Delta t \tag{1.4}$$

and none in Δt with probability

$$P_0(\Delta t) = P\{N(\Delta t) = 0\} = 1 - P\{N(t) = 1\}$$

$$= 1 - \lambda \Delta t \tag{1.5}$$

Here we have assumed that the probability of more than one arrival in Δt is negligible because Δt is chosen as sufficiently small. These probabilities are a direct consequence of the way we select Δt; that is, there is either no arrival or only one arrival in Δt. In calculating these probabilities, it is implicitly assumed that the arrival (input) process is stationary with constant arrival rate λ and the arrivals are orderly and mutually independent. The stationary property means that arrivals in $(t_0, t_0 + t)$ depend only on t but not on t_0.

Since, in each subinterval Δt, there can be no arrival or only one arrival, the event of an arrival in Δt can be regarded as a Bernoulli trial, which has only two possible outcomes. That is, for each trial, an arrival occurs in Δt with probability

$$p = P\{N(\Delta t) = 1\} = \lambda \frac{t}{n} \tag{1.6}$$

and no arrivals occur in Δt with probability

$$q = P\{N(\Delta t) = 0\} = 1 - \frac{\lambda t}{n} \tag{1.7}$$

Thus, the random variable $N(\Delta t)$ can be regarded as a Bernoulli random variable. Clearly, the total number of trials is n because $t = n \Delta t$. It follows that the probability of exactly k arrivals in t is given by the binomial distribution

$$P_k(t) = P\{N(t) = k\} = \binom{n}{k} p^k (1-p)^{n-k}, k = 0, 1, \dots, n,$$

$$\text{where } \binom{n}{k} = \frac{n!}{k! \, (n-k)!} \text{ is the binomial coefficient.}$$

The above binomial distribution can be rewritten as

$$P_k(t) = \frac{n!}{k! \, (n-k)!} \left[\frac{\lambda t}{n} \right]^k \left[1 - \frac{\lambda t}{n} \right]^{n-k}$$

$$= \frac{n(n-1) \dots (n-k+1)}{k! \, n^k} \frac{(\lambda t)^k}{\left[1 - \frac{\lambda t}{n} \right]^k} \left[1 - \frac{\lambda t}{n} \right]^n$$

As n $\to \infty$, this expression becomes

$$P_k(t) = \frac{(\lambda t)^k}{k!} e^{-\lambda t}, \quad k = 0, 1, 2, \ldots \tag{1.8}$$

since the following limits hold:

$$\lim_{n\to\infty} \frac{n(n-1)\ldots(n-k+1)}{n^k \left(1 - \frac{\lambda t}{n}\right)^k} = 1$$

and

$$\lim_{n\to\infty} \left(1 - \frac{\lambda t}{n}\right)^n = e^{-\lambda t}$$

In words, formula (1.8) states that the probability of exactly k arrivals in a period of length t is a Poisson distribution with a constant arrival rate λ.

We shall calculate the mean and variance of the Poisson random variable $N(t)$ using the formula (1.8). By definition, the mean value of $N(t)$ is the mathematical expectation of $N(t)$

$$E[N(t)] = \sum_{k=0}^{\infty} k\, P_k(t) = \sum_{k=0}^{\infty} k\, \frac{(\lambda t)^k}{k!} e^{-\lambda t}$$

$$= (\lambda t) e^{-\lambda t} \sum_{k=1}^{\infty} \frac{(\lambda t)^{k-1}}{(k-1)!}$$

$$= \lambda t \tag{1.9}$$

since the last summation equals $e^{\lambda t}$.

The variance of $N(t)$ by definition is given by

$$Var[N(t)] = E[N^2(t)] - (E[N(t)])^2$$

$$= E[N^2(t)] - (\lambda t)^2$$

Note that the second moment of N(t) is given by

$$E[N^2(t)] = \sum_{k=0}^{\infty} k^2 P_k(t) = \sum_{k=1}^{\infty} k^2 P_k(t)$$

$$= \sum_{k=1}^{\infty} k(k-1) P_k(t) + \sum_{k=1}^{\infty} k P_k(t)$$

$$= \sum_{k=2}^{\infty} k(k-1) \frac{(\lambda t)^k}{k!} e^{-\lambda t} + \lambda t$$

$$= (\lambda t)^2 e^{-\lambda t} \sum_{k=2}^{\infty} \frac{(\lambda t)^{k-2}}{(k-2)!} + \lambda t$$

$$= (\lambda t)^2 + \lambda t$$

where the last summation equals $e^{\lambda t}$. Hence, the variance of N(t) is simply

$$Var[N(t)] = \lambda t \qquad (1.10)$$

These results show that the mean and the variance of the Poisson random variable are equal. It is also stated that the Poisson distribution has equal mean and variance.

Example 1-1. During the busy hours of the weekday, in a certain telephone exchange, phone calls demanding services arrived are as follows:

> 4,542 calls on Monday
> 6,586 calls on Tuesday
> 7,698 calls on Wednesday
> 8,884 calls on Thursday
> 7,683 calls on Friday

What is the probability that a call arrives?

The fractional parts of call arrivals are:

$$\frac{4,542}{35,493} = 0.13 \text{ on Monday}$$

$$\frac{6,586}{35,493} = 0.19 \text{ on Tuesday}$$

$$\frac{7,698}{35,493} = 0.22 \text{ on Wednesday}$$

$$\frac{8,884}{35,493} = 0.25 \text{ on Thursday}$$

$$\frac{7,683}{35,493} = 0.22 \text{ on Friday.}$$

We see that the arithmetic average of the fractional parts for the individual day is close to the number 0.20. So, the probability sought under the given conditions is approximately 0.20. It appears that the fractional part of call arrivals under usual conditions will not deviate significantly from this number during various weekdays.

1.3 Superposition of Independent Poisson Processes

The Poisson input process representing the arrival of customers has a very important characteristic that the sum of independent Poisson input processes is also a Poisson process.

Consider two independent Poisson random variables $N_1(t)$ and $N_2(t)$ with arrival rates λ_1 and λ_2 respectively.

Let $\qquad N(t) = N_1(t) + N_2(t)$

Note that the event $\{N(t) = k\}$ is the sum or union of the independent events $\{N_1(t) = j, N_2(t) = k-j\}$ for $j = 0, 1, ..., k$. It follows that

$$P\{N(t) = k\} = \sum_{j=0}^{k} P\{N_1(t) = j, N_2(t) = k-j\}$$

$$= \sum_{j=0}^{k} P\{N_1(t) = j\} P\{N_2(t) = k-j\}$$

$$= \sum_{j=0}^{k} \frac{(\lambda_1 t)^j}{j!} e^{-\lambda_1 t} \frac{(\lambda_2 t)^{k-j}}{(k-j)!} e^{-\lambda_2 t}$$

$$= e^{-(\lambda_1+\lambda_2)t} \sum_{j=0}^{k} \frac{(\lambda_1 t)^j}{j!} \frac{(\lambda_2 t)^{k-j}}{(k-j)!}$$

$$= \frac{e^{-(\lambda_1+\lambda_2)t}}{k!} \sum_{j=0}^{k} \frac{k!}{j!(k-j)!} (\lambda_1 t)^j (\lambda_2 t)^{k-j}$$

$$= \frac{e^{-(\lambda_1+\lambda_2)t}}{k!} [(\lambda_1 + \lambda_2) t]^k$$

$$= \frac{(\lambda t)^k}{k!} e^{-\lambda t}, \quad k = 0, 1, \dots,$$

where $\lambda = \lambda_1 + \lambda_2$.

By induction, the above result is valid for the case of m independent Poisson processes with arrival rates $\lambda_1, \lambda_2, \dots, \lambda_m$ respectively and $\lambda = \lambda_1 + \lambda_2 + \dots + \lambda_m$.

Example 1-2. In certain suburb areas, a telephone exchange may be used by several suburb areas so that the total call input process to the telephone exchange is a combination of the individual suburb call arrival processes. In this case, we have

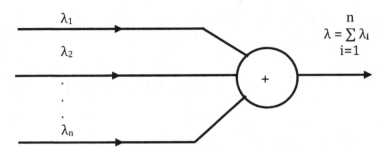

Fig. 1-3. The superposition of call arrival processes.

In practice, the individual call input process may not be Poisson. However, if the number of the individual input processes is large, then the total call input process may be assumed to be approximately Poisson.

1.4 Decomposition of a Poisson Process

Suppose that the arrivals of a Poisson process $N(t)$ with rate λ are divided into two processes $N_1(t)$ and $N_2(t)$ according to probabilities λ_1/λ and λ_2/λ, respectively, where $\lambda = \lambda_1 + \lambda_2$. We shall show that $N_1(t)$ and $N_2(t)$ are independent Poisson processes with rates λ_1 and λ_2, respectively. We shall make use of the fact that if A and B are independent events, then

$$P(AB) = P(A)\,P(B)$$

By the definition of conditional probability, we write

$$P\{N_1(t) = n_1, N_2(t) = n_2\} = P\{N_1(t) = n_1, N_2(t) = n_2 \mid N(t) = n\} \times P\{N(t) = n\}$$

Since

$$P\{N_1(t) = n_1, N_2(t) = n_2 \mid N(t) = n\} = \binom{n}{n_1}\left(\frac{\lambda_1}{\lambda}\right)^{n_1}\left(\frac{\lambda_2}{\lambda}\right)^{n-n_1}$$

where $n = n_1 + n_2$, $\lambda = \lambda_1 + \lambda_2$ and

$$P\{N(t) = n\} = \frac{(\lambda t)^n}{n!}\, e^{-\lambda t}$$

it follows that

$$P\{N_1(t) = n_1, N_2(t) = n_2\} = \frac{n!}{n_1!\,(n - n_1)!}\left[\frac{\lambda_1}{\lambda}\right]^{n_1}\left[\frac{\lambda_2}{\lambda}\right]^{n-n_1}\frac{(\lambda t)^n}{n!}\, e^{-\lambda t}$$

$$= \frac{(\lambda_1 t)^{n_1}}{n_1!}\, e^{-\lambda_1 t}\,\frac{(\lambda_2 t)^{n_2}}{n_2!}\, e^{-\lambda_2 t}$$

$$= P\{N_1(t) = n_1\}\, P\{N_2(t) = n_2\}$$

This expression states that the decomposition of a Poisson process into two random processes according to the probabilities λ_1/λ and λ_2/λ, results in two independent Poisson processes with rates λ_1 and λ_2, respectively. By the same reasoning, we deduce that the decomposition of a Poisson process with rate λ into m random processes according to the probabilities λ_i/λ, i = 1, 2, ... , m results in m independent Poisson processes with rates λ_1, λ_2, ..., λ_m, respectively, where $\lambda = \lambda_1 + \lambda_2 + ... + \lambda_m$.

Example 1-3. In a large city, telephone calls are handled by many exchanges. A portion of the Poisson input call process may go to different exchanges. In this case, we have

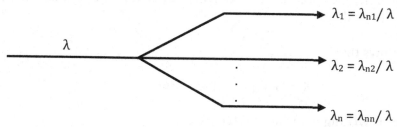

Fig. 1-4. The decomposition of a Poisson call arrival process.

We see that in practical situations, the probabilities for the decomposition of the call arrival process are governed naturally by the behavior of the customers, not by the telephone exchange.

1.5 The Exponential Interarrival Time Distribution

The interarrival times of an input process provides another useful mathematical representation for the specification of the arrival process.

Let T_i, i = 1, 2, ... , be the ith arrival epoch and $X_i = T_i - T_{i-1}$, with T_0 = 0, be the ith interarrival time.

Assume that the interarrival times are mutually independent and identically distributed with the common distribution function

$$F(t) = P\{X_i \leq t\}$$

Let

$$F_c(t) = 1 - F(t) = P\{X_i > t\}$$

Clearly, $F_c(t)$ is the probability that the interarrival time X_i is greater than t. Furthermore, assume that the input process is stationary and has an arrival rate λ. To determine $F_c(t)$, we consider the probability

$$P\{X_i > t+\Delta t\} = P\{X_i > t+\Delta t \mid X_i > t\} \; P\{X_i > t\}$$

If Δt is sufficiently small such that in Δt there can be only one arrival with probability $\lambda \Delta t$ and no arrival with probability 1- $\lambda \Delta t$, the probability of more than one arrival in Δt is extremely small and can be neglected.

It follows that

$$P\{X_i > t+\Delta t \mid X_i > t\} = 1 - \lambda \Delta t$$

Thus, rewriting in terms of $F_c(t)$, we get

$$F_c(t+\Delta t) = (1 - \lambda \Delta t) \; F_c(t)$$

or $\qquad \dfrac{F_c(t+\Delta t) - F_c(t)}{\Delta t} = \lambda F_c(t)$

Passing to the limit $\Delta t \rightarrow 0$ gives

$$\frac{d}{dt} F_c(t) = \lambda\, F_c(t)$$

which has the solution

$$F_c(t) = F_c(0)\, e^{-\lambda t} = e^{-\lambda t} \qquad\qquad (1.11)$$

where $F_c(0) = 1$. This result shows that under the given assumptions, the interarrival time of the input process is distributed in an exponential distribution.

The exponential distribution (1.11) has the mean or expected value

$$E[X_i] = \int_0^\infty t\, d\, F(t) = \int_0^\infty t\, \lambda\, e^{-\lambda t}\, dt$$

$$= \frac{1}{\lambda}$$

and the variance

$$Var[X_i] = E[X_i^2] - (E[X_i])^2$$

$$= \int_0^\infty t^2\, \lambda\, e^{-\lambda t}\, dt - \frac{1}{\lambda^2}$$

$$= \frac{1}{\lambda^2}$$

1.6 The Markov Property or Memoryless Property

A continuous non-negative random variable X is said to have the Markov property if for every $t > 0$ and every $x > 0$,

$$P\{X > t + x \mid X > t\} = P\{X > x\} \qquad\qquad (1.12)$$

A random variable with the Markov property is often said to have no memory. The Markov property plays an important role in the analysis of queueing systems.

To show that the exponential distribution (1.11) indeed has the Markov property, consider the conditional probability

$$P\{X > t + x \mid X > t\} = \frac{P\{X > t + x\}}{P\{X > t\}}$$

$$= \frac{e^{-\lambda(t+x)}}{e^{-\lambda t}}$$

$$= P\{X > x\}$$

which is (1.12).

The exponential distribution can be used to describe the service time distribution.

1.7 Relationship Between the Poisson Distribution and the Exponential Distribution

Recall that the input process of arrivals can be modeled either by N(t), the number of arrivals in t as a Poisson distribution in (1.8), or X(t), by the interarrival time as an exponential distribution in (1.11). Note that the following events are equivalent,

$$\{X_i > t\} \sim \{T_i - T_{i-1} > t\} \sim \{T_i > T_{i-1} + t\}$$

and that the event $\{T_i > t_{i-1} + t \mid T_{i-1} = t_{i-1}\}$ occurs if and only if the value of N(t) does not change in the period $(t_{i-1}, t_{i-1} + t)$. Thus,

$$\{T_i > t_{i-1} + t \mid T_{i-1} = t_{i-1}\} \sim \{N(t_{i-1} + t) - N(t_{i-1}) = 0\}$$

Since equivalent events have equal probability, then

$$P\{X_i > t \mid T_{i-1} = t_{i-1}\} = P\{N(t_{i-1} + t) - N(t_{i-1}) = 0\}$$

$$= P\{N(t) = 0\}$$

where $t_{i-1} = 0$ and $N(0) = 0$ have been used. This result shows that the conditional probability is independent of i and T_{i-1}. It follows that

$$P\{X_i > t \mid T_{i-1} = t_{i-1}\} \quad = P\{X_i > t\}$$

$$= P\{N(t) = 0\}$$

$$= e^{-\lambda t}$$

The equivalence of the event $\{N(t) = 0\}$ and the event $\{X_i > t\}$ is the direct consequence of the assumption that the input arrival process has stationary and independent increments. It can be seen from (1.8) and (1.11) that the Poisson distribution and the exponential distribution are equivalent. This is the reason why the input arrival process can be modeled either by a Poisson distribution or by an exponential distribution.

1.8 The Service Time Distribution

In the study of queueing systems, the service time is the duration of time for which a customer occupies a server for his service.

Let T denote the service time. In many applications, it is usually assumed that the service time has the exponential distribution

$$P\{T \le t\} = 1 - e^{-\mu t}, t \ge 0 \tag{1.13}$$

where μ is the service rate.

A typical example is the holding time of a telephone line that is the conversation duration a caller has made.

According to the measurement made by AT&T, the holding time distribution is close to exponential. A typical distribution of the holding time is shown in Fig. 1-5.

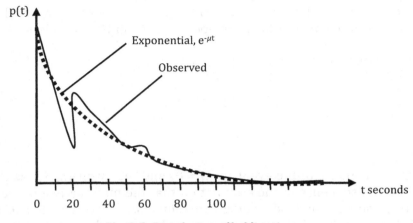

Fig. 1-5. Distribution of holding times.

From Fig. 1-5, we see that the observed curve follows an exponential distribution very closely, except for a dip in the holding time between 20 and 30 seconds. This dip corresponds to a time that is too short for a successful call and too long for a misdialed or unsuccessful call. In practice, relatively few calls have holding times between these two limits. Furthermore, the distribution of holding times of long-distance calls is also close to exponential. Therefore, for practical applications, in most cases, the holding times of telephone calls are very nearly exponential.

In the analysis of many queueing systems, the method used is valid only for systems with Poisson input and exponential service time. In fact, without these two assumptions, the analysis in general would be formidable and may encounter insurmountable difficulties.

Example 1-4. The input process of patient services at a clinic is usually arranged or controlled by appointments. In this case, the Poisson input model may not be a suitable representation because it is approximately regular and predictable. Although the Poisson input and exponential service time assumptions are widely used in the study of the performance of queueing systems due to their mathematical convenience, we must justify their applicability in practice. In the study of waiting time of patients at a clinic, the Poisson input model is not suitable and another model must be sought. If mathematical analysis is too difficult or formidable, computer simulation may be the approach to be employed.

1.9 The Residual Service Time Distribution

Consider an M/G/1 queueing system, where the arrival or input process is Poisson and the service time X assumes a common probability distribution function H(x) and probability density function h(x). That is,

$$H(x) = P\{ X \le x \}$$
and
$$h(x) = \frac{dH(x)}{dx}$$

In Fig. 1-6, it shows that the service time X starts at time t_1 and terminates at time t_2.

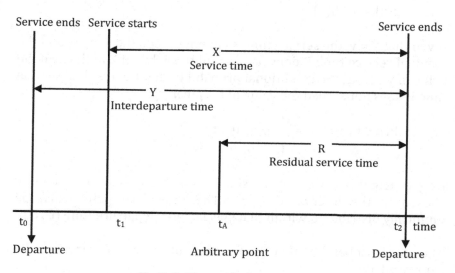

Fig. 1-6. The residual service time.

Let t_A be an arbitrarily chosen point in X, where $X = t_2 - t_1$ and $t_1 < t_A < t_2$. The time interval $R = t_2 - t_A$ is defined as the residual service time. To define R precisely, the arbitrary time point t_A must be in the time interval (t_1, t_2), which is the service time $X = t_2 - t_1$. However, this particular service time must be contained in a special interdeparture time $Y = t_2 - t_0$. The probability of occurrence of such interval Y is $g(y)$ dy, where $g(y)$ is the probability density function of Y.

Since the special interdeparture time Y contains the service time X, the above probability should be proportional to the length x, as well as the probability of occurrence of X; that is, $h(x)$ dx. Thus, we may write

$$g(x)\, dx = K\, x\, h(x)\, dx$$

where K is a constant that is determined to normalize the density function $g(x)$. Integrating the above expression from 0 to ∞ yields $K = 1/\tau = \mu$, where τ is the mean service time and μ is the service rate. It follows that

$$g(x) = \mu\, x\, h(x)$$

Given that $Y = y$, the service time X cannot exceed the value y and the residual service time R does not exceed the value of X and hence the value of y. Thus, the conditional probability that the residual service time R does not exceed the value of t is given by

$$P\{R \le t \mid Y = y\} = \begin{cases} t/y, & 0 \le t \le y \\ 0, & \text{otherwise} \end{cases}$$

Notice that $0 \le t \le y$ is true, since a point t_A has been randomly chosen in the interval (t_1, t_2). Therefore, this point must be uniformly distributed within that special interdeparture time (t_0, t_2).

The joint probability density function of R and Y can then be expressed as

$$P\{t \le R \le t + dt, y \le Y \le y + dy\} = \left(\frac{dt}{y}\right)\left[\mu\, y\, h(y)\, dy\right]$$

$$= \mu\, h(y)\, dy\, dt, \quad 0 \le t \le y$$

Let the probability density function of R be f(t) and the probability distribution function be F(t). Now integrating the above expression over y from t to ∞ gives the probability density function

$$f(t) = \mu \int_{t}^{\infty} h(y) \, dy = \mu \, [1 - H(t)]$$

and

$$F(t) = \mu \int_{0}^{t} [1 - H(y)] \, dy$$

The Laplace-Stieljes transform of F(t) is then given by

$$\hat{F}(s) = \frac{1}{\tau s} \, [1 - \hat{H}(s)]$$

from which we calculate the average residual service time

$$E[R] = - \frac{d\hat{F}(s)}{ds} \bigg|_{s=0}$$

$$= \tfrac{1}{2} \, \overline{X^2} / \overline{X} \tag{1.14}$$

1.10 The Birth and Death Process

The birth and death process is a special case of the discrete–state continuous–time Markov process, which is often called a continuous–time Markov chain. In application, it is usually convenient to describe Markov chains in terms of random variables.

Let N(t) be a random variable specifying the state (size) of a population at time t. The process N(t) is said to be a birth and death process if N(t) can take only non-negative integral values and can have positive and negative jumps. Furthermore, the process is said to be in state k at time t if N(t) = k, that is, the size of the population at time t is k.

For a complete description of a birth and death process, we assume $N(t)$ has the following properties:

(a) The probability of transition from state k to state k+1 in the time interval $(t, t+\Delta t)$ is $\lambda_k \Delta t + o(\Delta t)$, where λ_k is called the birth rate in state k and Δt is an infinitesimal quantity. $o(\Delta t)$ denotes all infinitesimal quantities small by comparison with Δt such that

$$\lim_{\Delta t \to 0} \frac{o(\Delta t)}{\Delta t} = 0$$

(b) The probability of transition from state k to state k-1 in Δt is $\mu_k \Delta t + o(\Delta t)$, where μ_k is called the death rate in state k;

(c) The probability of transition from state k to a state other than a neighboring state k-1 or k+1 in Δt is negligible, that is, if $| i - k | > 1$, then $P_{ik} (\Delta t) \approx 0$, where $P_{ik}(\Delta t)$ is the transition probability.

(d) The probability of no transitions in Δt is equal to $1 - (\lambda_k + \mu_k) \Delta t + o(\Delta t)$.

It follows from these properties and the formula of total probability that

$$P_k (t + \Delta t) = \sum_{i=0}^{\infty} P_i (t) P_{ik} (\Delta t)$$

$$= P_k(t) P_{kk} (\Delta t) + P_{k-1}(t) P_{k-1,k}(\Delta t) + P_{k+1}(t) P_{k+1,k}(\Delta t) + o(\Delta t)$$

$$= P_k(t) [1 - (\lambda_k + \mu_k) \Delta t] + \lambda_{k-1} P_{k-1}(t) \Delta t + \mu_{k+1} P_{k+1}(t) \Delta t + o(\Delta t)$$

or

$$\frac{P_k(t + \Delta t) - P_k(t)}{\Delta t} = - (\lambda_k + \mu_k) P_k(t) + \lambda_{k-1} P_{k-1}(t) + \mu_{k+1} P_{k+1}(t) + \frac{o(\Delta t)}{\Delta t}$$

As $\Delta t \to 0$, we have

$$\frac{d}{dt} P_k(t) = -(\lambda_k + \mu_k) P_k(t) + \lambda_{k-1} P_{k-1}(t) + \mu_{k+1} P_{k+1}(t), \quad k = 0, 1, 2, \ldots \quad (1.15)$$

with $\lambda_{-1} = \mu_0 = P_{-1}(t) = 0$.

This set of differential–difference equations represents the dynamic behavior of the birth and death process. In order to solve these equations for $P_k(t)$, the initial conditions $P_k(0)$ are required. It appears that no solution has been found yet.

Notice that the probabilities

$$P\{N(t) = k\} = P_k(t)$$

change in the course of time, and further depend on the initial conditions $P_i(0)$. Thus, it seems that the probabilities $P_k(t)$ can be defined only for given t and $P_i(0)$, $i \geq 0$. In application, however, it is usually regarded possible to speak of the probability p_k of finding a system in the state k, independent of the chosen moment t of time and of the initial conditions $P_i(0)$. To justify such a practice theoretically, one can attempt to establish that as $t \to \infty$, the process N(t) approximates without limit to a certain stationary process independent of the initial conditions $P_i(0)$. To be more concrete, it is necessary to establish that as $t \to \infty$, the probabilities $P_k(t)$ tend to certain constant numbers p_k independent of the initial conditions $P_i(0)$.

These numbers p_k may then be taken as probabilities for finding the birth and death process in one or other defined state, since, on the one hand, the probability p_k does not depend on the initial conditions $P_i(0)$, and, on the other hand, comes as near as needed to the real probability $P_k(t)$ if the birth and death process continues for a sufficiently long time.

Let us call a Markov process, characterized by the transition probabilities $P_{ik}(t)$, "transitive" if there exists such a $t > 0$ that $P_{ik}(t) > 0$ for all non-negative integral values i and k. Thus, for a transitive process there exists such a period of time in the course of which the change of a process from any state to any other state is possible. It is clear that the birth and death process N(t) is transitive.

According to Markov's theorem, which states that for any transitive Markov process $P_{ik}(t)$, the limit

$$\lim_{t \to 0} P_{ik}(t) = p_k, \ 0 \le i \le n, \ 0 \le k \le n$$

exists and does not depend on i, where n is a finite positive integer. Khinchin has proved that the existence of the above limit is also valid for Markov processes with infinite number of states, that is, the above limit exists for $i \ge 0$, $k \ge 0$ [3]. Therefore, Markov's theorem is extended to cover the birth and death process. It follows that if at the starting moment 0, the process is in state i, then the probability $P_k(t)$ is given by the formula of total probability

$$P_k(t) = \sum_{i=0}^{\infty} P_i(0) \, P_{ik}(t)$$

As $t \to \infty$,

$$\lim_{t \to \infty} P_k(t) = \lim_{t \to \infty} \sum_{i=0}^{\infty} P_i(0) \, P_{ik}(t)$$

$$= \sum_{i=0}^{\infty} P_i(0) \lim_{t \to \infty} P_{ik}(t)$$

$$= \sum_{i=0}^{\infty} P_i(0) \, p_k$$

$$= p_k$$

since $\sum_{i=0}^{\infty} P_i(0) = 1$.   Under this condition, the process is said to be in statistical equilibrium, or simply, in equilibrium.

Since, as $t \to \infty$, the right-hand sides of all equations (1.15) have limits, all derivatives on the left-hand sides tend to limits, but each of these limits can only be zero. If any derivative tends to other than zero, then, as $t \to \infty$, $|P_k(t)|$ would increase infinitely which is impossible on account of Markov's theorem and its extension. Thus, it may be concluded that

$$\lim_{t \to \infty} \frac{d}{dt} P_k(t) \to 0, \ k \ge 0.$$

Consequently, the set of equations in (1.15) in the limit as t → ∞ gives,

$$-(\lambda_k + \mu_k)p_k + \lambda_{k-1}\, p_{k-1} + \mu_{k+1}\, p_{k+1} = 0, \quad k \geq 0 \tag{1.16}$$

with $\lambda_{-1} = \mu_0 = p_{-1} = 0$.

This set of equations together with the normalization condition

$$\sum_{k=0}^{\infty} p_k = 1 \tag{1.17}$$

uniquely determines the state probabilities p_k. If we let

$$z_k = \lambda_{k-1}\, p_{k-1} - \mu_k\, p_k\,,$$

then (1.16) can be written in the form

$$z_k = z_{k+1}$$

since $z_0 = \lambda_{-1}\, p_{-1} - \mu_0\, p_0 = 0$, it follows that

$$z_k = 0, \quad k = 0, 1, 2, \ldots$$

and hence

$$p_k = \frac{\lambda_{k-1}}{\mu_k}\, p_{k-1}$$

Consequently,

$$p_k = \frac{\lambda_0\, \lambda_1\, \ldots\, \lambda_{k-1}}{\mu_1\, \mu_2\, \ldots\, \mu_k}\, p_0\,, \quad k = 1, 2, \ldots. \tag{1.18}$$

The probability p_0 is determined by the normalization condition (1.17)

$$p_0 = \frac{1}{1 + \sum\limits_{k=1}^{\infty} \dfrac{\lambda_0\, \lambda_1\, \cdots\, \lambda_{k-1}}{\mu_1\, \mu_2\, \cdots\, \mu_k}} \qquad (1.19)$$

where the infinite series on the right-hand side is assumed to be convergent.

In application, the birth and death process can be used to model queueing systems with Poisson input and exponential service time. This will be demonstrated in later chapters.

We have determined the state probabilities p_k for the birth and death process by first establishing the set of differential-difference equations (1.15). Then, by virtue of Markov's theorem, we obtained, in the limit as $t \rightarrow \infty$, the set of difference equations (1.16) and (1.17) from which the desired solution for p_k is determined in (1.18) and (1.19). However, it is instructive to explore another simpler approach known as the state-transition-rate diagram.

According to the properties (a) and (b), the state transition rates λ_k and μ_k are specified. We can draw a state-transition-rate diagram for the birth and death process. Let the circle with a number in it denote the state of the process and the curves with an arrow specify the transition rate entering a state or out of a state.

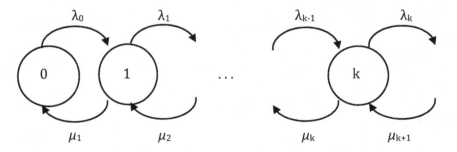

Fig. 1-7. State-transition-rate diagram for the birth and death process.

It is important to note that in the state-transition-rate diagram, the self-loop from a state back to the same state is not included, since such a diagram specifies only the rates of transition from a state to other (different) neighboring states. Thus, in constructing a state-transition-rate diagram only properties (a) and (b) are needed.

When the birth and death process is in statistical equilibrium, the transition rates entering a state and leaving the same state must be equal. This notion provides for us a simple means to write the equilibrium equations by observing the state-transition-rate diagram Fig. 1-7.

Transition rates into state $k = \lambda_{k-1} \, p_{k-1} + \mu_{k+1} \, p_{k+1}$

and

Transition rates out of state $k = (\lambda_k + \mu_k) \, p_k$.

Equating these two transition rates gives

$$\lambda_{k-1} \, p_{k-1} + \mu_{k+1} \, p_{k+1} = (\lambda_k + \mu_k) \, p_k \, , \quad k \geq 0$$

with $\lambda_{-1} = \mu_0 = p_{-1} = 0$. This is exactly the set of equations in (1.16).

1.11 The Outside Observer's Distribution and the Arriving Customer's Distribution

In the investigation of queueing systems with Poisson input, the state probability $\{p_k\}$ was derived from equations that relate the proportions of time that the system spends in various states to each other. Therefore, for systems in equilibrium, p_k may be interpreted either as the proportion of time that the system spends in state k or as the probability that the system is in state k at an arbitrary moment. The distribution $\{p_k\}$ of time that the system spends in each state is called the outside observer's distribution. Let $\{\pi_k\}$ denote the state probability distribution that an arriving customer would encounter; that is, the state probability distribution with respect to only those arrival instants of the customer. In queueing theory, the distribution $\{\pi_k\}$ is called the arriving customer's distribution. It is important to realize that, in general, the outside observer's distribution $\{p_k\}$ and the arriving customer's distribution $\{\pi_k\}$ are different. However, for queueing systems with Poisson input, these two distributions are identical. This fact is of central importance in the investigation of queueing systems with Poisson input. What follows will show this fact.

Suppose that n customers arrive in the time interval $(t_0, t_0 + T)$. Let T_k denote the total time that the system spends in state k. The proportion of time that the system spends in state k is

$$P_k = \frac{T_k}{T}$$

Another important quantity of practical significance is the proportion of arriving customers that finds the system in state k:

$$\pi_k = \frac{n_k}{n}$$

where n_k is the total number of arriving customers finding the system in state k in the time interval $(t_0, t_0 + T)$.

To find the relation between the outside observer's distribution and the arriving customer's distribution, let N(t) denote the state (the number of customers) in the system at time t and let $A(t, t + \Delta t)$ be the event that a customer arrives in the time interval $(t, t + \Delta t)$. Also, let

$$P_k(t) = P\{N(t) = k\}$$
$$= \text{the probability that the outside observer finds the system in state k at time t,}$$

and

$$\pi_k(t) = \text{the probability that the system is in state k at time t just prior to an arrival epoch.}$$

By definition, we write

$$\pi_k(t) = \lim_{\Delta t \to 0} P\{N(t) = k \mid A(t, t + \Delta t)\}$$

Using Baye's rule of probability theory, we can write

$$P\{N(t) = k \mid A(t, t + \Delta t)\} = \frac{P\{A(t, t + \Delta t) \mid N(t) = k\}\, P_k(t)}{\sum_{i=0}^{\infty} P\{A\{t, t + \Delta t) \mid N(t) = i\}\, P_i(t)}$$

It follows that

$$\pi_k(t) = \lim_{\Delta t \to 0} \frac{P\{A(t, t + \Delta t) \mid N(t) = k\} P_k(t)}{\sum_{i=0}^{\infty} P\{A\{(t, t + \Delta t) \mid N(t) = i\} P_i(t)}$$

If the arrival process described by $A(t, t + \Delta t)$ is a birth process with rate λ_k when the system is in state k, then

and
$$P\{A(t, t + \Delta t) \mid N(t) = k\} = \lambda_k \Delta t + o(\Delta t)$$

$$\pi_k(t) = \lim_{\Delta t \to 0} \frac{[\lambda_k \Delta t + o(\Delta t)] P_k(t)}{\sum_{i=0}^{\infty} [\lambda_i \Delta t + o(\Delta t)] P_i(t)}$$

$$= \frac{\lambda_k P_k(t)}{\sum_{i=0}^{\infty} \lambda_i P_i(t)}$$

This result shows that in general $\pi_k(t)$ and $P_k(t)$ are not equal. However, if the arrival process is Poisson with rate λ, then $\lambda_i = \lambda, i \geq 0$ and the summation in the denominator becomes one, so that

$$\pi_k(t) = P_k(t)$$

or when the system is in equilibrium, as $t \to \infty$, we have

$$\pi_k = p_k$$

This result implies that for a queueing system with a Poisson input, the arriving customer's distribution and the outside observer's distribution are equal. This relation is called Poisson Arrivals, See Time Averages (PASTA) [8].

It is important to point out that the arriving customer's distribution is needed for the calculation of the waiting time distribution for the queueing system.

Review

When the number of sources generating the traffic (the arrivals or input) to a queueing system is very large, the average arrival rate will be fairly constant. In this situation, the arrival process may be modeled by a Poisson process with a constant rate. We should remember that the random variable of the Poisson process is the number of arrivals in a given interval of time.

Another way to look at the arrival process is to consider the interarrival interval; that is, the length of the time interval between two consecutive arrival epochs, as a random variable. In this case, the arrival process is modeled by an exponential distribution. We have shown that these two models are equivalent mathematically.

The importance of the exponential distribution is that it has the memoryless property, which means that when the present behavior is known, the future behavior will be affected by the present behavior only and not the past history. The memoryless property facilitates the analysis of queueing systems. Without this memoryless property, the analysis may turn out to be difficult and formidable.

We have also shown that the residual service time distribution of an exponential distribution is the same as the original exponential distribution. This fact is very important for the investigation of waiting time distribution in a queueing system.

The birth and death process is a very powerful tool for modeling queueing systems. By choosing suitable birth and death rates, results obtained for the state probabilities of the birth and death process can be applied directly to the specific queueing system. This method will be used in the investigation of the Erlang loss and Erlang delay system in Chapter 2 and Chapter 3 respectively, as well as the Engset loss and Engset delay systems in Chapter 4 and some single server queueing systems in Chapter 5.

CHAPTER 2

QUEUEING SYSTEMS WITH LOSSES

2.1 Introduction

Having, in chapter 1, studied the properties of an input process of arriving customers, we shall in the following chapters investigate the fundamental questions connected with the service of this input process. Every queueing system must have several servers to serve the incoming customers. These servers can be very different according to the services they are to fulfill; in particular, a telephone operator, a shop assistant, a store cashier, or a doctor can be such a server. The process of service operates in such a way that each incoming customer occupies one of the servers that is free (unoccupied) at the moment of his arrival. While a server is busy with a customer, it is inaccessible to other incoming customers. The period of occupation of a server by one customer is called the service time.

A problem arises when an incoming customer finds all servers occupied (busy). In the one system (systems with losses), such a customer simply gets a refusal (or is lost) and the subsequent course of service continues as if this customer had not arrived at all. In the other kind of systems (systems allowing waiting in the queue), an incoming customer who finds all servers occupied, joins the queue and waits for a free server. These two types of systems differ from each other, not only in details of the solution of basic problems, but also in their very structure. The important fact is that the indices of the quality of service in these two types of systems are completely different. For a system with losses, the basic index is the probability of a refusal (loss). On the other hand, for a system with waiting, the main problem is the investigation of the waiting time as a random variable. Obviously, this problem has no meaning for a system with losses.

In view of the differences, we should investigate these two types of systems separately. We shall devote this chapter to systems with losses and the next chapter to systems allowing waiting.

2.2 The Erlang Loss System

In order to tackle the problems, we shall make the following assumptions.

(1)   We shall assume that all servers in a queueing system are always equally accessible to all incoming customers. Such a group is called a fully-accessible group of servers.

(2)   We shall always assume that for an input process, the number of customers arriving in a period of length t is distributed in a Poisson distribution with rate, λ.

(3)   For systems allowing waiting, the question of the order of service of waiting customers is of real significance in many problems. This service will be assumed first-come, first-served, or to be served in order of arrival.

(4)   The service time is an exponential distribution with service rate, μ; that is, the probability that the length of a service will be greater than t is equal to $e^{-\mu t}$. This assumption greatly facilitates necessary calculations in the investigation of problems of queueing systems.

The central part played by the exponential service time distribution is mainly due to the important property of memoryless. This means that for the exponential service time distribution, the law of distribution of the remaining parts of a service does not depend on its past history. It is precisely this memoryless property of the exponential distribution that simplifies, in most instances, the calculations.

A model for the queueing system with losses is depicted in Fig. 2-1, where the waiting capacity is zero; that is, no waiting is allowed.

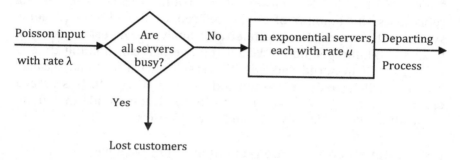

Fig. 2-1. Model for the queueing system with losses.

2.3 The Erlang Loss Formula

To determine the state probabilities p_k for our queueing system, we shall make use of the result obtained for the birth and death process by choosing the birth and death rates as follows:

$$\lambda_k = \begin{cases} \lambda, & k = 0, 1, \ldots, m-1 \\ 0, & k \geq m \end{cases} \tag{2.1}$$

and

$$\mu_k = \begin{cases} k\mu, & k = 0, 1, \ldots, m \\ 0, & k > m \end{cases} \tag{2.2}$$

Substituting the chosen λ_k and μ_k in (1.18) gives

$$p_k = \frac{a^k}{k!} p_0, \quad k = 1, 2, \ldots, m$$

where $a = \lambda/\mu$ is known as the offered traffic (or offered load) to the queueing system. It is numerically equal to the average number of customer arrivals (births) during an average holding time (an average length of life per person). In teletraffic theory, this dimensionless quantity, a, is internationally called erlang, named after A.K. Erlang, the father of teletraffic theory. One erlang represents the amount of traffic offered to the Erlang queueing system with losses. Thus, in teletraffic analysis, one erlang is equal to one call-hour per hour. When the average call holding time is expressed in hundred seconds, the resulting traffic unit is called hundred-call-seconds or centum-call seconds (CCS). In this case, 1 erlang equals 36 CCS, for there are 36 hundred-seconds in 1 hour. The traffic unit CCS is only used in North America.

By virtue of the normalization condition (1.17), we have

$$p_0 = \frac{1}{\sum_{j=0}^{m} \frac{a^j}{j!}} \qquad (2.3)$$

and hence

$$p_k = \frac{a^k/k!}{\sum_{j=0}^{m} \frac{a^j}{j!}}, \quad k = 0, 1, \ldots, m \qquad (2.4)$$

The average number of busy servers is called the carried traffic (or carried load):

$$a' = \sum_{k=0}^{m} k\,p_k = \sum_{k=1}^{m} \frac{a^k}{(k-1)!} \Big/ \sum_{j=0}^{m} \frac{a^j}{j!}$$

$$= a\,(1 - p_m) \qquad (2.5)$$

where $p_m = a^m\,p_0/m!$ \qquad (2.6)

and is the probability of an arriving customer finding all m servers occupied.

This is the probability of refusal that, in practice, is of particular significance, since in a queueing system with losses it is an indicator of the quality of the service. In teletraffic study, an arriving call that finds all lines occupied is known as call blocking. Thus, the blocking probability of a call equals the probability of refusal p_m.

The carried load per server is known as the server occupancy:

$$\rho = \frac{a'}{m}$$

(2.7)

The server occupancy is a measure of the degree of utilization of a group of servers and sometimes called the utilization factor.

In analysis, the value of λ and μ, as well as m, are assumed to be given. The performance of the queueing system with losses is measured by the probability p_m of loss. If the value of p_m is high, more servers may be needed. In general, the higher the value of m, the lower the value of p_m. On the contrary, the higher the value of λ or a, the higher the value of p_m. The probability distribution (2.4) is called the truncated Poisson distribution or Erlang's loss distribution. The probability of loss is given by (2.3) and (2.6)

$$p_m = B(m, a) = \frac{a^m / m!}{\sum_{k=0}^{m} a^k / k!}$$

(2.8)

In the United States, this expression is called the Erlang loss formula or the Erlang B formula and in Europe, it is called Erlang's first formula and is denoted by $E_{1, m}(a)$.

It should be pointed out that the truncated Poisson distribution (2.4) is valid for any service time distribution with a finite mean of $1/\mu$, even though the Markov property of the exponential service time distribution was used explicitly in the derivation. This remarkable fact was conjectured by Erlang himself in 1917, and

proved by Sevastyanov in 1957. Values for B(m, *a*) obtained from
(2.8) have been plotted against the offered traffic *a* in erlangs for
different values of the number m of servers [5].

More generally, the offered load generated by a Poisson input
process with a rate λ customers per unit time (in teletraffic study,
the unit time is usually taken as one (busy) hour) may be defined as
the average number of customer arrivals per holding time (in hours),
that is,

$$a = \int_0^\infty \lambda t \, dH(t) \, dt = \lambda \tau \tag{2.9}$$

where λt is the average number of customer arrivals in a fixed time
interval of t hours, H(t) is the holding time distribution and τ is the
average holding time.

Example 2-1. Consider a telephone exchange, which is operated as
an Erlang loss system. Ten thousand subscribers originate traffic at
an average rate of 1,500 calls per hour with a mean holding time of
200 seconds.

(a) What is the offered load to the system?

(b) How many erlangs of traffic is originated by a subscriber?

(c) What is the offered load in CCS per subscriber?

(d) If the exchange has a group of 100 trunk lines, what is the
blocking probability?

(e) What is the carried load per trunk line?

Since the number of subscribers is very large, the offered traffic may
be considered close to Poisson.

(a)   λ = 1,500 calls per hour

τ = 200 seconds = 1/18 hours

The offered traffic to the system is
a = λτ = 1,500 x 1/18 = 83.33 erlangs

= 1,500 calls per hour x 200/100 seconds
= 3,000 CCS

(b) The offered traffic per subscriber in erlang is equal to 83.33/ 10,000 = 0.0083 erlangs per subscriber.

(c) The offered load in CCS per subscriber is
3,000/ 10,000 CCS = 0.3 CCS per subscriber

(d) Since a = 83.33 erlangs
m = 100 trunk lines,

one may use the Erlang B curves [5], to find the blocking probability

$$B(m, a) = p_m$$
$$= 0.01$$

(e) The carried load is
$$a' = a\,[1 - B(m, a)]$$
$$= 83.33\,[1 - 0.01]$$
$$= 75 \text{ erlangs}$$

and the carried load per trunk line is

$$\rho = \frac{a'}{m} = 75/\,100 = 0.75 \text{ erlangs per trunk line.}$$

Example 2-2. Users (students) arrive at a computer room with 50 computers in a library at an average rate of 80 per hour. The average length of time using a terminal is 30 minutes. Users who find all computers occupied on arrival will leave.

(a) Assuming Poisson arrivals and exponential service time distribution, what is the probability that a student, who arrives at the computer room, finds all the computers occupied?

(b) What is the average number of computers occupied in the room?

(c) What is the computer occupancy?

Clearly, the operation of the computer room may be regarded as an Erlang queueing system with losses.

(a) The offered load is
$$a = \frac{\lambda}{\mu} = 80 \times 30/60 = 40 \text{ erlangs}$$

$$m = 50$$

Using the Erlang B curves [5], we find
$$p_{50} = B(50, 40) = 0.05$$

This is the probability that a student will find that, on arrival, all the computers are occupied.

(b) The average number of computers occupied is the carried load
$$a' = a(1 - p_m)$$
$$= 40(1 - 0.05)$$
$$= 38 \text{ erlangs}$$

(c) The carried load per computer is
$$\rho = \frac{a'}{m} = 38/50 = 0.76$$

The computer occupancy is high and the computers are well used.

Example 2-3. Students arrive at a library at an average rate of 10 per hour, demanding some information searches from a librarian. The average length of searching time is 30 minutes. There is a chair, which is used for a waiting student. Students who find both the librarian and the chair occupied will leave.

(a) Assuming Poisson arrivals and exponential searching times, what are the state probabilities

$$p_k = P\{N = k\}, \quad k = 0, 1, 2$$

for the system? Here, N is a random variable denoting the number of students in the system.

(b) What is the average number of students served per hour?

(c) If there are two librarians and no waiting chair, what is the average number of students served per hour?

We see that the situation (a) is not completely a loss system, so that the results obtained for the Erlang loss system do not apply. However, for the operation of this situation, it is still possible to make use of the concept of the birth and death process to solve our problem.

(a) Using the birth and death process, we choose the corresponding birth and death rates as follows,

$$\lambda_0 = \lambda_1 = 10 \text{ students per hour}, \lambda_2 = 0,$$

and $\quad \mu_1 = \mu_2 = 1/\tau = \dfrac{1}{30/60} = 2 \text{ students per hour}$

Applying (1.18) and (1.19), we find the state probabilities

$$p_0 = \dfrac{1}{1 + \dfrac{\lambda_0}{\mu_1} + \dfrac{\lambda_0 \lambda_1}{\mu_1 \mu_2}} = \dfrac{1}{1 + 10/2 + 10/2 \times 10/2} = 1/31$$

$$p_1 = \dfrac{\lambda_0}{\mu_1} p_0 = 5/31$$

$$p_2 = \dfrac{\lambda_0 \lambda_1}{\mu_1 \mu_2} p_0 = 25/31$$

(b) The average number of students served per hour is

$$\mu_0\, p_0 + \mu_1\, p_1 + \mu_2\, p_2 = 0 + 2 \times 5/31 + 2 \times 25/31$$

$$= 60/31$$

(c) Now this situation becomes an Erlang loss system with two servers. The corresponding birth and death rates are

$$\lambda_0 = \lambda_1 = 10,\ \lambda_2 = 0$$

and $\mu_0 = 0,\ \mu_1 = 2,\ \mu_2 = 4$

By virtue of (1.18) and (1.19), we find

$$p_0 = 2/37$$

$$p_1 = 10/37$$

$$p_2 = 25/37$$

Hence, the average number of students served per hour is

$$\mu_0\, p_0 + \mu_1\, p_1 + \mu_2\, p_2 = 0 + 2 \times 10/37 + 4 \times 25/37$$

$$= 3.24$$

Note that the above state probabilities can also be obtained by means of the formula (2.4) for the Erlang loss system with $\lambda = 10$ and $\mu = 2$.

Example 2-4. Consider the telephone trunking problem. Telephone exchange A is to serve subscribers in a nearby exchange B, as shown in Fig. 2-2. Suppose that during the busy hour, the subscribers generate a Poisson traffic with a rate of 600 calls per hour, which requires trunk lines to exchange B for an average holding time of 3 minutes. What is the number of trunk lines needed for a grade of service of 0.01? What is the loss traffic? Further, what is the utilization factor?

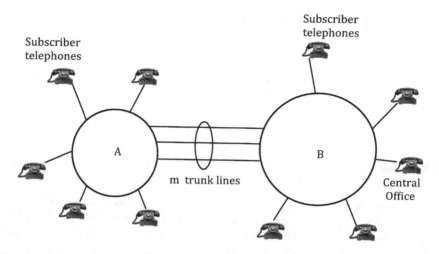

Fig. 2-2. Two nearby telephone exchanges.

Grade of service is a measure of congestion expressed as the probability that a call is blocked or delayed. This probability represents the percentage of the offered traffic, which is blocked or delayed during the busy hour. If the design of a system is based on the fraction of calls blocked (the blocking probability), then the system is said to be engineered on a blocking basis. Blocking criteria are often used for the dimensioning of interoffice trunk group.

Since $\lambda = 600$ calls/ hour
        $\tau = 3$ minutes
the offered traffic $a = \lambda \tau = 30$ erlangs.

Using the Erlang B curves [5], we find that for a grade of service 0.01, the number of trunk lines needed is 41.

The carried load is

$$a' = a\,[1 - B\,(m, a)]$$
$$= 30\,[1\text{-}0.01]$$
$$= 29.7 \text{ erlangs}$$

The loss traffic is equal to

$$a - a' = 0.3 \text{ erlangs}$$

The utilization factor is

$$\rho = \frac{a'}{m} = 0.72.$$

Review

The operation of queueing systems may be divided into two types. On the one hand, customers are not allowed to wait when all servers are busy on arrival. These customers are said to be blocked and will leave the queueing system as if they had not arrived. This type of queueing systems is known as the Erlang loss system. The performance index of this type of system is the probability of loss.

On the other hand, blocked customers are allowed to wait in a queue for a free server. This type of system is known as the Erlang delay system. The performance index of this type of system is the probability of waiting and the waiting time distribution. A.K. Erlang has solved the problem completely. However, in this chapter, we solved the problem using the birth and death process as a model. The approach employed is simple. We obtained the probability of loss and the carried load. The traffic loss is the difference of the offered load and carried load. If the traffic loss is high, this indicates that additional servers are needed for a given grade service.

Since the formulas involve summations and factorials, the calculation is formidable. In application, the Erlang curves are used from which we can read the result easily with no calculation at all [5].

CHAPTER 3

QUEUEING SYSTEMS ALLOWING WAITING

3.1 Introduction

In queueing systems allowing waiting, an arriving customer must wait for a service when and only when he finds all m servers occupied. Thus our former probability of loss (ie. the probability of finding all the servers occupied) can be called the probability of waiting. This probability plays only a part in the evaluation of the quality of service. However, for a queueing system with waiting, this part is comparatively small, since even if a significant majority of customers have to wait, the service must be regarded as completely satisfactory if the time of waiting is mostly very short. It is not the frequency of waiting which plays a decisive role, but the nature of the waiting time W as a random variable. Thus, the final objective in the investigation of queueing systems with waiting is always the determination of the distribution function of the waiting time W.

3.2 The Erlang Delay System

Consider a queueing system with a Poisson input with rate λ; that is; customers arriving in the system follow a Poisson process with rate λ. Customers will be served by m servers. The service times are assumed to be mutually independent and exponentially distributed with a mean $1/\mu$. The queue discipline is first-come, first-served or customers are served in order of their arrival. A queueing model is shown in Fig. 3-1.

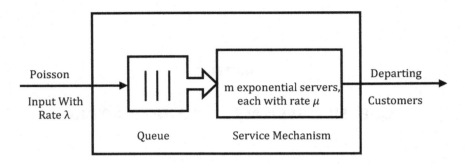

Fig. 3-1. Queueing model for the Erlang delay system.

Note that if the state of the system is k and is less than m, that is, k < m, the system is operated in exactly the same way as the Erlang loss system.   In this case, all k customers are served and no customers are waiting. However, if k ≥ m, all m servers are occupied and there are k–m customers waiting in the queue. Further, the change from state m to state m+1 is now possible. Taking the waiting customers into account, the Erlang delay system can be modeled by the birth and death process with the following choice of birth and death rates:

$$\lambda_k = \lambda, \ k = 0, 1, \ldots \tag{3.1}$$

and

$$\mu_k = \begin{cases} k\mu, \ k < m \\ \\ m\mu, \ k \geq m \end{cases} \tag{3.2}$$

Under equilibrium conditions, the state probability distribution p_k can be obtained by substituting these birth and death rates into (1.18) and (1.19) to yield:

For k < m,

$$p_k = \frac{a^k}{k!} p_0, \ k < m \tag{3.3}$$

For $k \geq m$,

$$p_k = \frac{a^k}{m! \, m^{k-m}} \, p_0, \quad k \geq m \tag{3.4}$$

Using the normalization condition

$$\sum_{k=0}^{\infty} p_k = 1$$

it follows that

$$\frac{1}{p_0} = \sum_{k=0}^{m-1} \frac{a^k}{k!} + \frac{m^m}{m!} \sum_{k=m}^{\infty} \left(\frac{a}{m}\right)^k$$

$$= \sum_{k=0}^{m-1} \frac{a^k}{k!} + \frac{a^m}{m!} \frac{1}{1 - a/m} \tag{3.5}$$

where the condition $\rho = a/m < 1$ has been assumed. Theoretically this condition is needed for the convergence of the infinite series. It also has a physical meaning that the server occupancy ρ cannot be greater or equal to 1.

Further, the probability of finding all m servers occupied (the probability of waiting) is, in consequence of (3.4), equal to

$$p_w = \sum_{k=m}^{\infty} \pi_k$$

$$= \sum_{k=m}^{\infty} p_k$$

$$= \frac{a^m}{m!} \frac{p_0}{1 - a/m} \tag{3.6}$$

where p_k and p_0 are given by (3.4) and (3.5) respectively.

It should be pointed out that a full solution for the problem has been given by Erlang [6]. In the United States, the probability of waiting in (3.6) denoted by

$$p_w = C\,(m,a) = P\{W > 0\}$$

is called the Erlang delay formula or the Erlang C formula. In Europe, it is known as Erlang's second formula and is denoted by $E_{2,m}\,(a)$. Values for $C\,(m,a)$ obtained from (3.6) have also been plotted against the offered traffic a in erlangs for different values of m, known as the Erlang C curves [5].

The average number of customers in the Erlang delay system is given by

$$E[N] = \sum_{k=0}^{\infty} k\,p_k$$

$$= \sum_{k=0}^{m} k\,\frac{a^k}{k!}\,p_0 + \sum_{k=m+1}^{m} \frac{k}{m!}\,\frac{a^k}{m^{k-m}}\,p_0$$

The first sum on the right-hand side is equal to

$$a \sum_{k=0}^{m-1} \frac{a^{k-1}}{(k-1)!}\,p_0 = a[\,1 - C\,(m,a)\,]$$

and the second sum can be written as

$$\frac{m^{m-1}\,a\,p_0}{m!}\left(\sum_{k=m+1}^{\infty} k\,\rho^{k-1}\right) = \rho\,C\,(m,a)\left(\frac{m+1-a}{1-\rho}\right)$$

since the quantity in the brackets on the left-hand side is simply equal to

$$\frac{d}{d\rho} \sum_{k=m+1}^{\infty} \rho^k = \frac{d}{d\rho}\left(\frac{\rho^{m+1}}{1-\rho}\right)$$

$$= \rho^m\,\frac{(m+1-m\rho)}{(1-\rho)^2}$$

where $\rho = a/m < 1$.

Consequently, we find the average number of customers in the Erlang delay system

$$E[N] = a\left[1 - C(m,a)\right] + \rho\, C(m,a)\, \frac{m + 1 - a}{1 - \rho}$$

$$= \frac{\rho}{1 - \rho}\, C(m,a) + m\,\rho \qquad\qquad (3.7)$$

Note that the first term on the right-hand side is the average queue length (the average number of waiting customers in the queue) $E[N_q]$ and the second term is the average number of busy servers. Furthermore, using Little's formula (3.22) or expression (3.23), we can calculate the mean waiting time without knowing the distribution function of the waiting time:

$$E[W] = \frac{E[N_q]}{\lambda} = \frac{C(m,a)}{m\,\mu\,(1 - \rho)} \qquad\qquad (3.8)$$

Here our notations W and N_q are random variables, while in Little's formula (3.22) and expression (3.23) W and N_q represent their mean values.

Comments. By definition, the carried load is the average number of busy lines, that is,

$$a' = \sum_{k=1}^{m-1} k\, p_k + m \sum_{k=m}^{\infty} p_k$$

$$= \sum_{k=1}^{m-1} \frac{a^k}{(k-1)!}\, p_0 + m \sum_{k=m}^{\infty} p_k$$

$$= a\left(1 - p_{m-1} - \sum_{k=m}^{\infty} p_k\right) + m \sum_{k=m}^{\infty} p_k$$

$$= a - a\, p_{m-1} + (m - a) \sum_{k=m}^{\infty} p_k$$

Using (3.4) and (3.6), the last term on the right-hand side can be written as

$$(m\text{-}a) \ \frac{a^m}{m!} \ \frac{p_0}{1 - a/m} = a \, p_{m-1}$$

Consequently, we obtain

$$a' = a$$

This result shows that for an Erlang C (or delay) system, the carried load is always equal to the offered load. This is intuitively clear, because, in a queueing system allowing waiting, all calls are served and there is no loss of traffic.

3.3 The Distribution Function of the Waiting Time

Now we shall determine the probability $P\{W > t\}$ that a customer arriving in the Erlang delay system at a randomly chosen moment has a waiting time greater than t. Suppose that this customer has found the system in state k. Let $P_k\{W > t\}$ denote the conditional probability of the waiting time greater than t on the assumption that the system is in state k. Using the formula of total probability

$$P\{W > t\} = \sum_{k=0}^{\infty} p_k \, P_k\{W > t\}$$

or, since $P_k\{W > t\} = 0$ for $k < m$ and $t > 0$, there is no waiting at all.

$$P\{W > t\} = \sum_{k=m}^{\infty} p_k \, P_k\{W > t\} \qquad (3.9)$$

The state probabilities p_k are known and are given by (3.3), (3.4) and (3.5). It remains to find the conditional probabilities $P_k\{W > t\}$ for all $k \geq m$.

Note that there are k–m waiting customers when the said customer arrives in the system. This customer will be served after the (k–m+1)th completion of service during the time period t. In other words, he will not be served in the time period t if there are k–m or fewer completions of service. Let r = k–m and $q_r(t)$ be the probability of this event. Then,

$$P_k \{W > t\} = \sum_{r=0}^{k-m} q_r(t), k \geq m.$$

But the process of completion of service, during the time that the test customer is waiting, obeys an exponential distribution with service rate mμ, since the probability that no completion will occur during time t from the moment when all servers are occupied is equal to

$$(e^{-\mu t})^m = e^{-m\mu t}$$

Thus, it follows from Section 1.7, the number of completions when all m servers are occupied is subject to the Poisson distribution

$$q_r(t) = \frac{(m\mu t)^r}{r!} e^{-m\mu t}, \ 0 \leq r \leq k-m$$

Hence,

$$P_k \{W > t\} = \sum_{r=0}^{k-m} \frac{(m\mu t)^r}{r!} e^{-m\mu t}, \ k \geq m$$

Substituting this result $P_k \{W > t\}$ and the state probabilities p_k of (3.4) and (3.5) into (3.6) gives

$$P\{W > t\} = \sum_{k=m}^{\infty} p_k \sum_{r=0}^{k-m} \frac{(m\mu t)^r}{r!} e^{-m\mu t}$$

$$= \sum_{k=m}^{\infty} \left(\frac{a}{m}\right)^{k-m} p_m \sum_{r=0}^{k-m} \frac{(m\mu t)^r}{r!} e^{-m\mu t}$$

$$= p_m \, e^{-m\mu t} \sum_{r=0}^{\infty} \frac{(m\mu t)^r}{r!} \sum_{k=m+r}^{\infty} \left(\frac{a}{m}\right)^{k-m}$$

$$= p_m \, e^{-m\mu t} \sum_{r=0}^{\infty} \frac{(m\mu t a)^r}{m^r \, r!} \sum_{k=m+r}^{\infty} \left(\frac{a}{m}\right)^{k-m-r}$$

$$= p_m \, e^{-m\mu t} \sum_{r=0}^{\infty} \frac{(\lambda t)^r}{r!} \, \frac{1}{1 - a/m}$$

$$= \frac{p_m}{1 - a/m} \, e^{-(m\mu - \lambda)t}$$

$$= p_w \, e^{-(m\mu - \lambda)t}, \; t \geq 0 \qquad\qquad (3.10)$$

where $\;p_w = \dfrac{p_m}{1 - a/m}$

This result shows that under the given conditions the waiting time follows the exponential distribution with rate $m\mu - \lambda$ and the probability of waiting is

$$P\{W > 0\} = p_w = \frac{m}{m - a} \, p_m \qquad\qquad (3.11)$$

where p_m is given by (3.3) or (3.4) and (3.5).

Comments. It is instructive to present a simple method for the calculation of the waiting time distribution function for the M/M/m queueing system [9].

Recall that the state probabilities p_k in (3.4), (3.5) and (3.6) are the probabilities that the system is in state k at a randomly chosen moment. More precisely, in the calculation of the waiting time distribution function $P\{W > t\}$, the probabilities π_k that the system is in state k just prior to the arrival epoch of the customer should be used. In other words, instead of (3.9), we should use the following expression

$$P\{W > t\} = \sum_{k=m}^{\infty} \pi_k \; P_k\{W > t\} \qquad (3.12)$$

where $\{\pi_k\}$ for $k \geq 0$ is known as the arriving customer's distribution and $\{p_k\}$ as the outside observer's distribution. For Poisson arrival process, the arriving customer's distribution and the outside observer's distribution are equal [8], that is,

$$\pi_k = p_k, \; k \geq 0 \qquad (3.13)$$

Equation (3.13) is known as the PASTA (Poisson Arrivals, See Time Averages) property [8].

Note that during the waiting time t, all m servers must be busy. Any one of the m busy servers can contribute a service rate μ. The resultant service rate of the m servers as a group is essentially m μ. In other words, the whole group of m busy servers acts like a single server with an exponential service time of rate m μ. At this point, k–m+1 customers are waiting for service in the system, and each of them will take an exponential service time distribution of mean $1/m\mu$. It follows from the memoryless property of the exponential distribution that the waiting time of the test customer is composed of k–m+1 exponential service times, each of which has an exponential distribution with rate $m\mu$. Thus, the conditional waiting time W^*_{k-m+1} is

$$W^*_{k-m+1} = R + \sum_{j=1}^{k-m} W_j, \; k \geq m \qquad (3.14)$$

where R is the residual (exponential) service time of the group of m customers being served on the arrival of the test customer, which has the same distribution as W_j because of the Markov property. W_j is the waiting time of the jth customer in the queue, which is equal to the service time of the group of m customers. Clearly, the conditional waiting time W^*_{k-m+1} is the sum of k–m+1 identical exponential random variables each with rate $m\mu$ and hence can be regarded as an Erlang random variable of order $k-m+1$ with parameter (rate) $m\mu$, whose probability density function is

$$e_{k-m+1}(x) = \left[\frac{(m\mu)^{k-m+1}x^{k-m}}{(k-m)!}\right]e^{-m\mu x}, \ k \geq m, x \geq 0 \qquad (3.15)$$

Then the conditional waiting time distribution is simply

$$P_k\{W > t\} = \int_t^\infty e_{k-m+1}(x)\,dx$$

$$= \int_t^\infty \left[\frac{(m\mu)^{k-m+1}x^{k-m}}{(k-m)!}\right]e^{-m\mu x}\,dx \qquad (3.16)$$

Substituting (3.3), (3.4), and (3.16) into (3.12) results in the waiting time distribution function

$$P\{W > t\} = \sum_{k=m}^\infty \left[\frac{a^k}{m!\,m^{k-m}}\right]p_0 \int_t^\infty \left[\frac{(m\mu)^{k-m+1}x^{k-m}}{(k-m)!}\right]e^{-m\mu x}\,dx$$

$$= m\mu\, p_m \int_t^\infty e^{-(m\mu-\lambda)x}\,dx$$

$$= \left[\frac{m\,p_m}{m-a}\right]e^{-(m\mu-\lambda)t}, t \geq 0 \qquad (3.17)$$

which is the desired solution (3.10).

It should be recognized that the simplicity of the calculation of the conditional waiting distribution results from the fact that, during waiting, the group of m busy servers acts like a single exponential server with rate $m\mu$ and the exponential service time has the memoryless property.

3.4 Little's Formula

Consider a queueing system allowing waiting. The input process and the service time distribution function are general, or arbitrary with no specific distribution, and independent. There are m servers. Further, the queue discipline or the order of service is also arbitrary.

Let $\Lambda(t)$ denote the number of arrivals in the system at time t and $\nabla(t)$ the number of departures.

Number of arrivals $\Lambda(t)$

Number of departures $\nabla(t)$

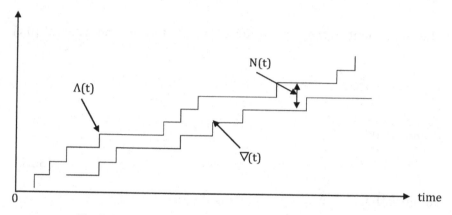

Fig. 3-2. Arrival process $\Lambda(t)$ and departure process $\nabla(t)$.

Then the number of customers in the system at an arbitrary time t is given by

$$N(t) = \Lambda(t) - \nabla(t)$$

The total area between the two curves of $\Lambda(t)$ and $\nabla(t)$ up to t represents the total time that the customers have spent in the system (measured in units of customer-seconds) in the time interval $(0, t)$.

Let this cumulative area be

$$A(t) = \int_{0}^{t} N(x)\, dx \qquad\qquad (3.18)$$

Then the average number of customers in the system is

$$N_t = \frac{A(t)}{t} \qquad (3.19)$$

Further, the average arrival rate in the time interval $(0, t)$ is

$$\Lambda_t = \frac{\Lambda(t)}{t} \qquad (3.20)$$

The time spent in the system per customer in the time interval $(0,t)$ is

$$T_t = \frac{A(t)}{\Lambda(t)}$$

It follows that

$$N_t = \Lambda_t \, T_t \qquad (3.21)$$

Suppose that as $t \to \infty$, the following limits exist

$$\lim_{t \to \infty} \Lambda_t = \lambda$$

and

$$\lim_{t \to \infty} T_t = T$$

Then the limit

$$\lim_{t \to \infty} N_t = N$$

also exists.

Taking the limits in (3.21) gives

$$N = \lambda \, T \qquad (3.22)$$

This result is known as Little's formula. It states that the average number of customers in a queueing system is equal to the product of the average arrival rate and the average time a customer spent in the system.

If Little's formula is applied to the queue only (not to the service mechanism), it becomes

$$N_q = \lambda W \tag{3.23}$$

where N_q is the average number of customers waiting in the queue and W is the average waiting time.

If Little's formula is applied to the service mechanism only, it becomes

$$N_s = \lambda / \mu \tag{3.24}$$

where N_s is the average number of customers in the service mechanism (average number of busy servers) and $1/\mu$ is the average service time. Clearly N_s is equal to the offered load a.

Since the total time a customer spent in the queueing system is equal to the sum of the waiting time in the queue and the service time, then their mean values satisfy the relation

$$T = W + 1/\mu \tag{3.25}$$

Multiplying this expression by λ gives

$$N = N_q + N_s \tag{3.26}$$

These formulas are useful for the calculation of the mean queue length N_q and the mean waiting time W when the average number of customers N in the queueing system is known.

Example 3-1. Consider the Erlang C system or the queueing system allowing waiting. If, on arrival, a call (or a customer) finds all lines (or servers) busy (or occupied), the call will wait for a free line. In this case, the grade of service is the probability of waiting.

Suppose that the call input process is Poisson with a rate of 600 calls per hour and the average holding time is 3 minutes.

(a) What is the number of lines needed for a grade of service 0.01?
(b) What is the average number of calls in the system?
(c) What is the mean waiting time?

Since in the Erlang C system, there is no loss of traffic, the carried load and the offered load are equal, that is,

$$a' = a$$

(a) The offered load is
$$a = \lambda \tau = 600 \ \frac{\text{calls}}{\text{hour}} \ \times \ 3 \ \text{minutes} \ \times \ \frac{1 \ \text{hour}}{60 \ \text{minutes}}$$

$$= 30 \ \text{erlangs}$$

Using the Erlang C curves [5], we find that for a grade of service (or probability of waiting) 0.01, the number of lines needed is 44.

(b) Since
$$\rho = a'/m = a/m = 30/44 = 0.68,$$

the average number of calls in the system is then given by

(3.7)
$$E[N] = \frac{\rho}{1-\rho} \ C\,(m, a) \ + \ m\rho$$

$$= 30.02 \ \text{calls}$$

(c) The mean waiting time is given by (3.8)
$$E[W] = \frac{C\,(m, a)}{m\,\mu\,(1-\rho)} = 0.13 \ \text{seconds}$$

Example 3-2. If the waiting time distribution is known, it is possible to obtain the mean waiting time $E[W]$ in (3.8) by means of (3.17), that is,

$$P\{W > t\} = \underline{\frac{m\ p_m}{m - a}}\ e^{-(m\mu - \lambda)t};\ t \geq 0$$

$$= p_w\ e^{-(m\mu - \lambda)t}$$

then $\quad E[W] = \int_0^\infty t\, d\left(p_w\, e^{-(m\mu - \lambda)t}\right)$

$$= \frac{p_w}{m\,\mu\,(1 - \rho)}$$

$$= \frac{C\,(m,\,a)}{m\,\mu\,(1 - \rho)}$$

which is (3.8), our desired solution.

Example 3-3. Consider an Erlang delay system with a Poisson input process with a rate λ calls per hour, m servers and exponential service times with mean $\tau = 1/\mu$ hours. Suppose that a test call arriving at the system at $t = 0$ finds all m servers busy with n waiting calls. Waiting calls are served in order of arrival. Service times are assumed to be mutually independent, identical and exponentially distributed random variables.

(a) What is the mean waiting time for the test call in the queue?
(b) Suppose that no new calls arrive at the system after $t = 0$. What is the expected length of time from the arrival of the test call at $t = 0$ until all the calls in the system complete their services?
(c) Let X be a random variable representing the order of completing the service of the test call. The event $X = k$ denotes the test call completing the service after $t = 0$. What is the probability $P\{X = k\}$?
(d) What is the probability that the test call has been completely served before the call immediately ahead of it in the queue?

As pointed out in Section 3.3, the group of m busy exponential lines behaves like a single line with rate $m\,\mu$. The test call must wait in the queue until n+1 completions of calls occur.

(a) The mean waiting time of the test call is simply equal to $(n + 1)/m\mu$.

(b) Note that all m lines must be busy (or occupied) when the test call is waiting. When the test call is being served, the expected length of completion of calls will vary according to the number of busy lines. Hence, the expected length of completion of calls is

$$\frac{1}{m\mu} + \frac{1}{(m-1)\mu} + \ldots + \frac{1}{\mu}$$

Therefore, the expected length of the time until the test completion of service of the test call is

$$\frac{n+1}{m\mu} + \frac{1}{\mu}\sum_{k=1}^{m}\frac{1}{k}$$

(c) Since n+1 calls must complete services before the test call enters service, we have

$$P\{X = k\} = 0 \text{ for } k \le n+1$$

For $n+2 \le k \le m+n+1$, the test call to complete service is equally likely to be any one of the m busy lines (by symmetry and the Markov property of the service time). Thus,

$$P\{X = k\} = 1/m \text{ for } k = n+2, n+3, \ldots , m+n+1$$

(d) Let A be the test call and B be the call immediately ahead of A. Then A completes service before B if A enters service before B finishes. The probability of this event is $(m-1)/m$. Once A is being served, A finishes before B has the probability $1/2$. Hence

$$P\{A \text{ finishes before } B\} = (m-1)/2m.$$

Review

For the Erlang delay system, the important performance indices are the probability of waiting and the mean waiting time. If the values of these indices are high, it indicates that additional servers are needed. Like the Erlang loss system, we can also use the Erlang delay curves to find the answer without any calculation [5].

Furthermore, mean value and higher moments of the waiting time can be obtained from the waiting time distribution.

Based on the observation that during the entire waiting period, the system operates as if the service is given by all the servers as a group so that the service rate becomes the number of servers multiplied by the service rate of the individual server. It turns out that the waiting time becomes a sum of exponential service times and hence has an Erlang distribution.

This fact greatly simplifies the calculation of the waiting time distribution function.

CHAPTER 4

THE ENGSET LOSS AND DELAY SYSTEMS

4.1 Introduction

In the study of the Poisson input process, we have assumed the arrival rate to be a constant. However, if the sources generating the input (customers) are finite, the arrival rate is not constant and the input process can no longer be modeled by the Poisson process.

If an input process is generated by a finite source which generates at a rate when idle and no generation when busy, this input process is called a quasi-random input.

Suppose that customer arrivals are generated by n independent sources, each with a rate γ when idle and zero when busy. In this chapter we shall assume the customer arrival process to be a quasi-random input process.

4.2 The Engset Loss System

Consider a queueing system with a quasi-random input generated by n identical, independent sources and served by s servers, where s < n. The service time distribution is exponential with mean $1/\mu$. An arriving customer who finds the system in state s will leave the system. A queueing system just described is known as an Engset loss system. A queueing model is shown in Fig. 4-1.

59

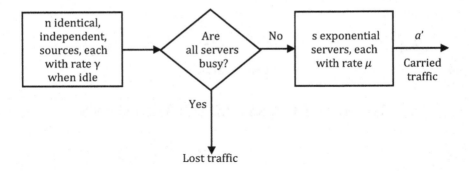

Fig. 4-1. Queueing model for the Engset loss system.

The assumption of quasi-random input implies that only idle sources can generate customers at rate γ. The probability of an arriving customer in the interval (t, t + Δt) when the system is in state k at time t is assumed to be

$$(n - k)\, \gamma\, \Delta t + o(\Delta t),\ 0 \le k \le s{-}1$$

and zero for k = s. It follows that the transition probability for the system to change from state k to state k+1 in (t, t + Δt) is given by

$$P_{k,\,k+1}\,(\Delta t) = \begin{cases} (n - k)\, \gamma\, \Delta t + o(\Delta t),\ 0 \le k \le s{-}1 \\ 0,\ k = s \end{cases}$$

Based on the assumption of exponential service times with mean 1/μ, the transition probability for the system to change from state k to state k − 1 in (t, t + Δt) is

$$P_{k,\,k-1}\,(\Delta t) = k\,\mu\,\Delta t + o(\Delta t),\ 0 \le k \le s$$

Further, the probability that the system will not change state in (t, t + Δt) is equal to

$$P_{k,\,k}\,(\Delta t) = \begin{cases} 1 - (n - k)\, \mu\, \Delta t + o(\Delta t),\ 0 \le k < s \\ \\ 1 - s\,\mu\,\Delta t + o(\Delta t),\ k = s \end{cases}$$

Like the Erlang systems, these state transition probabilities satisfy all the four properties of the birth and death process discussed in Section 1.10. Thus, the Engset loss system can be modeled by the birth and death process with the following birth and death rates:

$$\lambda_k = \begin{cases} (n-k)\,\gamma, & 0 \le k \le s-1 \\ 0, & k = s \end{cases}$$

and

$$\mu_k = k\,\mu, \quad 0 \le k \le s$$

Under equilibrium conditions, the state probabilities $p_k[n]$ of the Engset loss system can be obtained simply by substituting these birth and death rates into (1.18) to yield

$$p_k[n] = \binom{n}{k} \hat{a}^k\, p_0[n], \quad 0 \le k \le s$$

where $\hat{a} = \gamma/\mu$, is called the offered load per idle source. Using the normalization condition, we find

$$p_0[n] = \frac{1}{\displaystyle\sum_{k=0}^{s} \binom{n}{k} \hat{a}^k}$$

Hence,

$$p_k[n] = \frac{\displaystyle\binom{n}{k} \hat{a}^k}{\displaystyle\sum_{i=0}^{s} \binom{n}{i} \hat{a}^i} \tag{4.1}$$

If we introduce the new notation

$$p = \frac{\hat{a}}{1 + \hat{a}}$$

then

$$\hat{a} = \frac{p}{1 - p}$$

Substituting this $\hat{a}$ into the state probabilities expression and multiplying the numerator and denominator by $(1 - p)^n$ yields the truncated binomial distribution:

$$p_k[n] = \frac{\binom{n}{k} p^k (1 - p)^{n-k}}{\sum_{j=0}^{s} \binom{n}{j} p^j (1 - p)^{n-j}}, \quad 0 \leq k \leq s \tag{4.2}$$

Note that $p_1[1] = p$. Thus, p can be interpreted as the probability that an arbitrary idle source will be busy.

4.3 The Arriving Customer's Distribution for the Engset Loss System

Suppose that the system is in equilibrium. Let t_k be the portion of the observation time t in which the system is in state k. The average number of incoming customers during the time interval t_k is then equal to

$$(n - k) \gamma t_k = (n - k) \gamma t t_k/ t = (n - k) \gamma t p_k[n]$$

The proportion of incoming customers during t finding the system in state k is then equal to

$$\frac{(n-k)\,\gamma\,t\,p_k[n]}{\displaystyle\sum_{j=0}^{s}(n-j)\,\gamma\,t\,p_j[n]} = \frac{(n-k)\,p_k[n]}{\displaystyle\sum_{j=0}^{s}(n-j)\,p_j[n]}\,,\ 0 \le k \le s$$

which is just the probability that an arriving customer finds the system in state k:

$$\pi_k[n] = \frac{(n-k)\,p_k[n]}{\displaystyle\sum_{j=0}^{s}(n-j)\,p_j[n]}\,,\ 0 \le k \le s$$

Substitution of $p_k[n]$ and $p_j[n]$ in (4.1) into this expression gives

$$\pi_k[n] = \frac{(n-k)\dbinom{n}{k}\hat{a}^k}{\displaystyle\sum_{j=0}^{s}(n-j)\dbinom{n}{j}\hat{a}^j}$$

$$= \frac{\dbinom{n-1}{k}\hat{a}^k}{\displaystyle\sum_{j=0}^{s}\dbinom{n-1}{j}\hat{a}^j}$$

$$= p_k[n-1],\ 0 \le k \le s$$

The probability of a customer refusal is then given by

$$\pi_s[n] = \frac{\binom{n-1}{s} \hat{a}^s}{\sum_{j=0}^{s} \binom{n-1}{j} \hat{a}^j} , \ \hat{a} = \gamma/\mu \tag{4.3}$$

This formula is often called the Engset loss formula.

It is interesting to note that in the Engset loss system, the arriving customer's distribution with n sources is equal to the outside observer's distribution with n−1 sources; that is,

$$\pi_k[n] = p_k[n-1], \ 0 \le k \le s$$

This expression exemplifies that in general, the arriving customer's distribution is not equal to the outside observer's distribution. In a finite source system, the proportion of arriving customers finding all servers busy is always smaller than $p_s[n]$ because fewer arrivals occur during periods when all servers are busy. Values for $\pi_s[n]$ obtained from the Engset loss formula have been plotted against the offered load $\hat{a}$ per idle source for different values of the number of s servers (see for example, T.C. Fry [10]).

4.4 The Offered Load and Carried Load in the Engset Loss System

Consider the n-source Engset loss system with s servers. Suppose that the system is in equilibrium. Let N denote the average number of customers in the system. By definition, the carried load is

$$a' = N = \sum_{k=0}^{s} k \, p_k[n] = a \, (1 - \pi_s[n])$$

Since the average number of idle sources is equal to n−N, then the offered load, by definition, is equal to

$$a = (n - N) \, \gamma/\mu = \hat{a} \, (n - N)$$

$$= \frac{n\,\hat{a}}{1 + \hat{a}\,(1 - \pi_s\,[n])}$$

If $n = s$, then $\pi_s[n] = 0$ and the truncated binomial distribution (4.2) becomes the binomial distribution

$$p_k[n] = \binom{n}{k} p^k\,(1 - p)^{n-k},\ 0 \le k \le n.$$

4.5 The Engset Delay System

If arriving customers in the Engset system are allowed to wait in the queue when all servers are busy, then the birth and death rates must be modified as:

$$\lambda_k = (n - k)\,\gamma,\ 0 \le k \le n$$

and

$$\mu_k = \begin{cases} k\mu,\, 0 \le k \le s-1 \\[2mm] s\mu,\, s \le k \le n \end{cases}$$

Under equilibrium conditions, we find for $k \le s$, the situation is exactly the same as that of the Engset loss system. However, for $k \ge s$, the death rate will remain constant $s\mu$. Therefore, we now would have

$$P_k[n] = \begin{cases} \binom{n}{k} \hat{a}^k\,p_0[n],\ 0 \le k \le s-1 \\[3mm] \dfrac{n!\,\hat{a}^k\,p_0[n]}{(n-k)!\,s!\,s^{k-s}},\ s \le k \le n \end{cases}$$

where

$$1/p_0[n] = \sum_{k=0}^{s} \binom{n}{k} \hat{a}^k + \sum_{k=s+1}^{n} \frac{n!\,\hat{a}^k}{(n-k)!\,s!\,s^{k-s}}$$

Note that like the Engset loss system, the relation

$$\pi_k[n] = p_k[n-1], \ 0 \le k \le n$$

remains true for the Engset delay system.

The probability of waiting is

$$P\{W > 0\} = \sum_{k=s}^{n-1} \pi_k[n]$$

$$= \frac{(n-1)!}{s!} \, â^s \left(\frac{â}{s}\right)^{n-1-s} \sum_{k=0}^{n-1-s} \frac{1}{k!} \left(\frac{s}{â}\right)^k p_0[n-1] \quad (4.4)$$

4.6 The Waiting Time Distribution Function for the Engset Delay System

To calculate the waiting time distribution function, we use the expression

$$P\{W > t\} = \sum_{k=s}^{n} \pi_k[n] \, P_k\{W > t\} \quad (4.5)$$

$$= \sum_{k=s}^{n-1} p_k[n-1] \, P_k\{W > t\}$$

Since the probability $p_k[n-1]$ is known and the conditional probability of waiting is given by (see Section 3.3):

$$P_k\{W > t\} = e^{-s\mu t} \sum_{j=0}^{k-s} \frac{(s\mu t)^j}{j!}, \ s \le k \le n-1$$

It follows then that

$$P\{W > t\} = \sum_{k=s}^{n-1} \frac{(n-1)! \; \hat{a}^k}{(n-1-k)! \; s! \; s^{k-s}} \; p_0 \, [n-1] \; e^{-s\mu t} \sum_{r=0}^{k-s} \frac{(s\mu t)^r}{r!}$$

$$= p_0[n-1]\frac{(n-1)!}{s!} \, \hat{a}^s \, e^{-s\mu t}\left(\frac{\hat{a}}{s}\right)^{n-1-s} \times \left[\sum_{j=0}^{n-1-s} \sum_{r=j}^{n-1-s} \frac{(s\mu t)^{r-j}}{(r-j)!}\left(\frac{s}{\hat{a}}\right)^j \frac{1}{j!}\right]$$

The quantity in the square brackets can be written in the form

$$\sum_{r=0}^{n-1-s} \frac{1}{r!} \sum_{j=0}^{r}\binom{r}{j}(s\mu t)^{r-j}\left(\frac{s}{\hat{a}}\right)^j = \sum_{j=0}^{n-1-s}\frac{1}{r!}\left(s\mu t + \frac{s}{\hat{a}}\right)^j$$

Consequently, we find the waiting time distribution function for the Engset delay system

$$P\{W > t\} = c[n] \sum_{r=0}^{n-1-s} \frac{[\emptyset(t)]^r}{r!} \; e^{-s\mu t} \tag{4.6}$$

where

$$\emptyset(t) = s\mu\left(\frac{1}{\gamma} + t\right)$$

and

$$c[n] = p_0 \, [n-1] \, \frac{(n-1)!}{s!} \, \hat{a}^s \left(\frac{\hat{a}}{s}\right)^{n-1-s}$$

4.7 The Mean Waiting Time in the Engset Delay System

Now we calculate the conditional mean waiting time

$$E[\,W \mid N = k\,] = -\int_0^\infty t \, d \, P_k\{W > t\}$$

$$= - \sum_{r=0}^{k-s} \frac{1}{r! \, s\mu} [r\, r! - (r+1)!]$$

$$= \frac{k-s+1}{s\,\mu}, \quad s \le k \le n-1$$

The mean waiting time can then be written as

$$E[W] = \sum_{k=s}^{n-1} E\,[W \mid N = k]\, \pi_k\,[n]$$

$$= \sum_{k=s}^{n-1} \left(\frac{k-s+1}{s\,\mu} \right) \pi_k\,[n]$$

$$= \sum_{k=0}^{n-s-1} \frac{k+1}{s\,\mu}\, p_{s+k}\,[n-1] \tag{4.7}$$

4.8 The Offered Load and Carried Load in the Engset Delay System

For the Engset delay system, we assume that the size of the waiting room or the queue capacity is n–s so that no customer is lost. Let N denote the mean number of customers in the system and λ the mean arrival rate. It follows from Little's formula

$$N = \lambda\,W + a$$

However, the offered load is

$$a = (n - N)\, \hat{a}$$

and

$$\lambda = \mu\, a$$

Eliminating N and λ yields the offered load

$$a = \frac{n\,\hat{a}}{1 + \gamma\,(1/\mu + W)} \tag{4.8}$$

which is also equal to the carried load because no customer gets refusal. Note that W here denotes the mean waiting time due to the change of notation in Little's formula.

Comments. In the calculation of the probabilities $p_s[n]$ and $\pi_s[n]$, known as time congestion and call congestion in telephone industries, the binomial coefficient

$$\binom{n}{k} = \frac{n!}{k!\,(n-k)!}$$

is involved. It contains factorials. In order to compute the factorial, Stirling's formula is often used to obtain excellent approximation:

$$n! = (2\,\pi\,n)^{1/2}\,n^n\,e^{-n}$$

However, a more accurate formula may be used:

$$n! = (2\,\pi\,n)^{1/2}\,n^n\,e^{-n}\left(1 + \frac{1}{12n} + \frac{1}{288n^2} - \frac{139}{51840n^3} + \ldots\right)$$

Furthermore, the mean delay time in the Engset delay system is the sum of the mean waiting time and the mean service time; that is,

$$T = W + \tau$$

$$= \frac{n}{\lambda} - \frac{1}{\gamma} \tag{4.9}$$

Where (4.8) has been used for W. It is important to note that here $\lambda = (n - N)\,\gamma$ is the mean arrival rate to the Engset delay system; that is, the mean arrival rate of the quasi-random input process.

Expression (4.9) is useful for computer system performance evaluation. In this application, n sources or users request the use of the computer to process transactions. The time spent by the user in generating a request is called the think time. In this instance, $1/\gamma$ is the mean think time. The time duration from the instant a user generates a request until the computer completes the transaction is called the response time. Hence T is the mean response time. The rate λ in (4.9) at which transactions are processed is called the throughput.

Review

When the number of sources generating the input traffic is not large, the quasi-random input process must be used.

Both the Engset loss and Engset delay systems can also be modeled by the birth and death process with appropriate birth and death rates. As usual, the calculations of the performance indices are extremely complicated and curves must be used. Fry provides a good set of curves [10].

CHAPTER 5

QUEUEING SYSTEMS WITH A SINGLE SERVER

5.1 Introduction

Without the assumption of an exponential distribution for the service times, the investigation of systems allowing waiting is beset with great difficulties. One can succeed in getting simple results only in certain particular situations. From a practical point of view, the case of systems with a single server is of special importance. For these systems, the investigation of the waiting time can be simplified.

In the present chapter, the single server queue with Poisson input and general service time distribution will be considered. Further, the cases with server vacations and priority discipline are investigated. The case of single server queue with general independent input and exponential service times is also studied.

5.2 The M/M/1 Queue

The M/M/1 queue is characterized by a Poisson input process with rate λ, exponential service times with mean $1/\mu$, a single server and infinite waiting capacity. The state transitions in the M/M/1 queue satisfy all the four properties of the birth and death process stated in Section 1.10. We shall determine the state probabilities p_k, the mean $E[N]$ and the variance $Var[N]$ of the number of customers and the mean waiting time in the system.

For the M/M/1 queue, we have

$$\lambda_k = \lambda, \ \ k = 0, 1, \ldots$$

and

$$\mu_k = \mu, \ k = 1, 2, \ldots$$

Using these birth and death rates in (1.18) and (1.19) or (3.4) and (3.5) directly, we have

$$p_k = a^k p_0 = \rho^k p_0, \ k = 0, 1, \ldots \tag{5.1}$$

and

$$p_0 = 1 - a = 1 - \rho \tag{5.2}$$

where $\rho = a = \dfrac{\lambda}{\mu}$.

The probability of waiting is given by

$$p_w = \sum_{k=1}^{\infty} p_k$$
$$= \rho \tag{5.3}$$

The waiting time distribution function can be found easily from (3.4), (3.5), and (3.17) for m = 1 to be

$$P\{W > t\} = \rho \, e^{-\mu(1-\rho)t}, \ t \geq 0 \tag{5.4}$$

The average number of customers in the system is given by

$$E[N] = \sum_{k=0}^{\infty} k \, p_k = (1 - \rho) \sum_{k=0}^{\infty} k \rho^k$$

$$= \rho(1-\rho) \sum_{k=0}^{\infty} k \rho^{k-1}$$

$$= \rho(1-\rho) \frac{d}{d\rho} \left(\sum_{k=0}^{\infty} \rho^k \right)$$

$$= \frac{\rho}{1-\rho} \tag{5.5}$$

The behavior of $E[N]$ is plotted in Fig. 5-1.

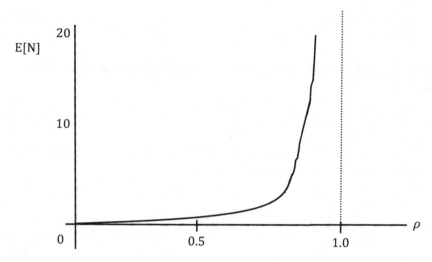

Fig. 5-1. Average number of customers E[N] vs. offered load ρ.

The variance Var[N] is given by

$$\text{Var}[N] = \sum_{k=0}^{\infty} k^2\, p_k - (\,E[N]\,)^2$$

The second moment of N is equal to

$$\sum_{k=0}^{\infty} k\,(k-1)\, p_k + E[N]$$

$$= \rho^2\,(1-\rho)\,\frac{d^2}{d\rho^2} \sum_{k=0}^{\infty} \rho^k + E[N]$$

$$= \frac{2\,\rho^2}{(1-\rho)^2} + \frac{\rho}{1-\rho}$$

$$= \frac{\rho^2 + \rho}{(1-\rho)^2}$$

Consequently, we have

$$\text{Var}[N] = \frac{\rho^2 + \rho}{(1-\rho)^2} - \frac{\rho^2}{(1-\rho)^2}$$

$$= \frac{\rho}{(1-\rho)^2} \tag{5.6}$$

Using Little's formula (3.22), we have the average time spent in the system

$$T = \frac{\rho}{\lambda\,(1-\rho)} = \frac{1}{\mu\,(1-\rho)} \tag{5.7}$$

and the mean waiting time W

$$W = T - \frac{1}{\mu}$$

$$= \frac{\rho}{\mu\,(1-\rho)} \tag{5.8}$$

As can be seen from (5.7) and (5.8), both T and W will grow in an unbounded fashion as $\rho \to 1$. This type of behavior is characteristic of almost all queueing systems.

5.3 The M/G/1 Queue and the Pollaczek-Khinchin Formula for the Mean Waiting Time

As mentioned previously, if the service time is not exponential, the method we used to investigate the birth and death process is no longer valid. In this section, we shall investigate the M/G/1 queue using the method of the imbedded Markov chain, which was introduced by D.G. Kendall in 1951 [1]. We shall call a time epoch a regeneration point of a random process if the Markov property holds at that point (epoch). Thus, at a regeneration point, the future evolution of the random process will depend only on the state at that regeneration point. It follows that every time point of a continuous-time Markov chain is a regeneration point. An imbedded Markov chain is a Markov chain whose states are defined at a discrete set of regeneration points that are imbedded in the non-negative time points.

We shall derive a formula known as the Pollaczek-Khinchin formula for the mean queue length using the method of imbedded Markov chain. Let N_k denote the total number of customers in the system right after the service completion epoch of the kth departing customer and X_k denote the number of customers that arrive during the service time of this departing customer. The idea is to try to find an equation relating the random variables N_k and N_{k+1}. Depending on the value of N_k, there are two cases as shown in Fig. 5-2(a) and Fig. 5-2(b).

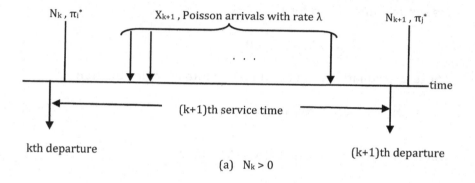

(a) $N_k > 0$

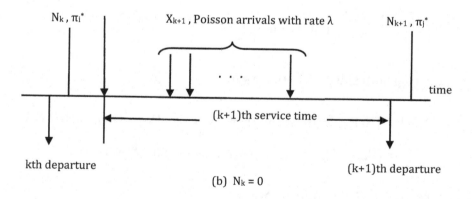

(b) $N_k = 0$

Fig. 5-2. Events occurring between two consecutive regeneration points (a) $N_k > 0$ and (b) $N_k = 0$.

If $N_k > 0$, then the (k+1)th departure will leave behind those same customers in the queue left by the kth customer except for himself, plus all those customers who arrived during the service time of the (k+1)th customer. Thus, we have

$$N_{k+1} = N_k - 1 + X_{k+1} , \quad N_k > 0.$$

If $N_k = 0$, then the queue is empty (no customer waits in the queue and the (k+1)th customer is served immediately on arrival) and the (k+1)th departure will leave behind those customers during his service time. Therefore,

$$N_{k+1} = X_{k+1} , \quad N_k = 0.$$

These two equations can be combined into a single equation

$$N_{k+1} = N_k - \delta(N_k) + X_{k+1} \qquad (5.9)$$

where the random variable $\delta(N_k)$ is defined as

$$\delta(N_k) = \begin{cases} 0, \text{ if } N_k = 0 \\ \\ 1, \text{ if } N_k > 0 \end{cases}$$

Squaring both sides of (5.9), we get

$$(N_{k+1})^2 = (N_k)^2 + \delta^2(N_k) + (X_{k+1})^2 + 2N_k X_{k+1} - 2X_{k+1}\delta(N_k) - 2N_k\delta(N_k)$$

$$= (N_k)^2 + \delta(N_k) + (X_{k+1})^2 + 2N_k X_{k+1} - 2X_{k+1}\delta(N_k) - 2N_k$$

since $\delta^2(N_k) = \delta(N_k)$ and $N_k\delta(N_k) = N_k$.

Taking expected values and using the fact that N_k and X_{k+1} are independent random variables, we have

$$E[(N_{k+1})^2] = E[(N_k)^2] + E[\delta(N_k)] + E[(X_{k+1})^2] + 2\ E[N_k]\ E[X_{k+1}] - $$
$$2\ E[X_{k+1}]\ E[\delta(N_k)] - 2\ E[N_k] \qquad (5.10)$$

Similarly, taking expected values of (5.9) yields

$$E[N_{k+1}] = E[N_k] - E[\delta(N_k)] + E[X_{k+1}] \qquad (5.11)$$

Assuming that statistical equilibrium conditions exist, as $k \to \infty$,

$$\lim_{k \to \infty} E[N_{k+1}] = \lim_{k \to \infty} E[N_k] = E[N]$$

$$\lim_{k \to \infty} E[(N_{k+1})^2] = \lim_{k \to \infty} E[(N_k)^2] = E[N^2]$$

$$\lim_{k \to \infty} E[X_{k+1}] = \lim_{k \to \infty} E[X_k] = E[X]$$

and

$$\lim_{k \to \infty} E[(X_{k+1})^2] = E[X^2]$$

It follows from (5.10) and (5.11) that

$$E[\delta(N)] = E[X]$$

and

$$E[N] = \frac{E[X]\ (1 - 2\ E[X]) + E[X^2]}{2\ (1 - E[X])} \qquad (5.12)$$

It remains to calculate the quantities $E[X]$ and $E[X^2]$. Note that $E[X]$ is simply the average number of customer arrivals during a service time and $E[X^2]$ is its second moment. Thus,

$$E[X] = \int_0^\infty \sum_{k=0}^\infty k\ \frac{(\lambda t)^k}{k!}\ e^{-\lambda t}\ dH(t)$$

$$= \lambda \int_0^\infty t \, dH(t)$$

$$= \lambda \tau$$

where τ is the average service time and $H(t)$ is the distribution function of the service time. Further,

$$E[X^2] = \int_0^\infty \sum_{k=0}^\infty k^2 \frac{(\lambda t)^k}{k!} e^{-\lambda t} \, dH(t)$$

$$= \int_0^\infty (\lambda^2 t^2 + \lambda t) \, dH(t)$$

$$= \lambda^2 (\delta^2 + \tau^2) + \lambda \tau$$

where δ^2 is the variance of the service time and the relation

$$\delta^2 = \int_0^\infty t^2 \, dH(t) - \tau^2$$

has been used.

Substituting the above two quantities into (5.12) gives

$$E[N] = \frac{\rho^2 (1 + \delta^2 / \tau^2)}{2 (1 - \rho)} + \rho \tag{5.13}$$

where $\rho = \lambda \tau$.

Since the average number $E[N]$ of customers in the system is equal to the average number $E[N_q]$ of customers in the queue plus the average number ρ of customers being served (busy servers),

$$E[N] = E[N_q] + \rho \tag{5.14}$$

Comparison of (5.13) and (5.14) gives a formula for calculating the mean queue length

$$E[N_q] = \frac{\rho^2}{2(1-\rho)}\left(1 + \frac{\delta^2}{\tau^2}\right) \qquad (5.15)$$

This is known as the Pollaczek-Khinchin formula for the mean queue length. Further, using Little's formula

$$E[N_q] = \lambda E[W]$$

where, instead of N_q and W as in (3.23), the notations of $E[N_q]$ and $E[W]$ are used for the mean queue length and the mean waiting time, respectively, the mean waiting time is

$$E[W] = \frac{\rho\tau}{2(1-\rho)}\left(1 + \frac{\delta^2}{\tau^2}\right) \qquad (5.16)$$

which is known as the Pollaczek-Khinchin formula for the mean waiting time.

In particular, when the service time is a constant equal to τ, the variance δ^2 of the constant service time becomes zero and hence

$$E[W] = \frac{\rho\tau}{2(1-\rho)}$$

which yields the smallest value of $E[W]$ in an M/G/1 queue. A single server queue with a Poisson input and constant service times is often referred to as an M/D/1 queue.

Table 5-1 shows some important expected values for a single-server queue with Poisson input, FIFO service rule, and various types of service time, where the constant $C_v = \delta/\tau$ is the coefficient of variation, and where τ and δ are respectively the mean and standard deviation of the service time T_s.

Expected Value	General T_s	Exponential T_s	Constant T_s
	$C_v = \delta/\tau$	$C_v = 1$	$C_v = 0$
$E[N_q]$	$\dfrac{\rho^2}{2(1-\rho)}(1+C_v^2)$	$\dfrac{\rho^2}{1-\rho}$	$\dfrac{\rho^2}{2(1-\rho)}$
$E[N] = E[N_q] + \rho$	$\dfrac{\rho^2}{2(1-\rho)}(1+C_v^2)+\rho$	$\dfrac{\rho}{1-\rho}$	$\dfrac{\rho^2}{2(1-\rho)}+\rho$
$E[W] = E[N_q]/\lambda$	$\dfrac{\rho\tau}{2(1-\rho)}(1+C_v^2)$	$\dfrac{\rho\tau}{1-\rho}$	$\dfrac{\rho t_s}{2(1-\rho)}$
$E[T] = E[W] + \tau$	$\dfrac{\rho\tau}{2(1-\rho)}(1+C_v^2)+\tau$	$\dfrac{\tau}{1-\rho}$	$\dfrac{(2-\rho)}{2(1-\rho)}t_s$

Table 5-1. Expected values for the M/G/1 queue.

Now we shall present a direct derivation of the Pollaczek-Khinchin formula for the mean waiting time in the M/G/1 queue. We have obtained the Pollaczek-Khinchin formula for the mean waiting time in (5.16) by using the method of the imbedded Markov chain. It is instructive to show that the Pollaczek-Khinchin formula can also be derived by means of the concept of residual service time discussed in Section 1.9.

Consider the M/G/1 queue with the first-come, first-served queue discipline or service in the order of arrival. We assume that the service times X_k, $k = 1, 2, \ldots$ are mutually independent, identically distributed and independent of the interarrival times. We also assume that the following time average and statistical average are equal:

$$\overline{X} = E[X_k] = \tau = 1/\mu = \text{mean service time}$$

$$\overline{X^2} = E[(X_k)^2] = \text{second moment of service time}$$

$$\overline{R} = E[R] = \text{mean residual service time.}$$

Now let us define the following events:

> B = on arrival, the test customer finds the server busy
> with probability $\rho = \lambda/\mu$ and N_q customers waiting;
> I = on arrival, the test customer finds the server idle with
> probability $(1 - \rho)$.

The mean waiting time of the test customer can then be expressed as

$$E[W] = \rho \, E[W \mid B] + (1 - \rho) \, E[W \mid I] \qquad (5.17)$$

Since there is no waiting when the event I occurs, the second term is always zero. Thus,

$$E[W \mid I] = 0.$$

Figure 5-3 illustrates the mean waiting time for a test customer in a M/G/1 queue.

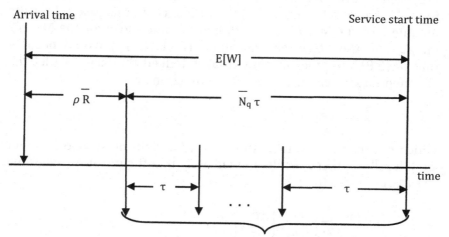

Arrival time

Service start time

$E[W]$

$\rho \bar{R}$ $\bar{N_q}\,\tau$

time

τ τ

. . .

Departures process, each with mean waiting time τ

Fig. 5-3. Mean waiting time in the M/G/1 queue.

When the event B occurs, the waiting time is composed of one residual service time R and a random number N_q of service times X. It follows that the conditional mean waiting is given by

$$E[W \mid B] = E[(R + X_1 + X_2 + \dots + X_{N_q}) \mid B]$$

$$= E[R] + \overline{N_q}\, \tau/\rho$$

$$= E[R] + E[W]$$

Using this result in (5.17) yields

$$E[W] = \frac{\rho\, E[R]}{1-\rho}$$

Finally, applying E[R] of (1.14) to this expression gives the desired Pollaczek-Khinchin formula (5.16) for the mean waiting time.

Comments. If the service time has an exponential distribution, then the M/G/1 queue becomes the M/M/1 queue. In this case, all results obtained for the M/G/1 queue should become those for the M/M/1 queue, and the Pollaczek-Khinchin formula (5.16) would become that of (5.8). Now let us show that is indeed the fact. We let the exponential service time distribution function be

$$H(t) = 1 - e^{-\mu t}$$

which has the mean service time $\tau = 1/\mu$ and the second moment $2\tau^2$. It follows these results immediately that the mean waiting time in (5.16) reduces to

$$E[W] = \frac{\rho\, \tau}{2(1-\rho)} \left(\frac{2\tau^2}{\tau^2} \right)$$

$$= \frac{\rho}{\mu\,(1-\rho)}$$

which is the mean waiting for the M/M/1 queue (5.8).

5.4 The M/G/1 Queue with Vacations

Observe that in a single-server queueing system, the server will have a busy period when there is at least one customer in the system, which is followed by an idle period when there is no customer present. A busy period is a time interval that begins when an arriving customer finds the system empty (with no customer present) and ends when a departing customer leaves the system empty. An idle period is the time interval between two successive busy periods. Thus, busy and idle periods occur alternately and form a cycle.

In the present section, we shall determine the mean waiting time in the M/G/1 queue on the assumption that at the end of each busy period, the server can go on vacation for a random interval of time with first moment $\overline{V}$ and the second moment $\overline{V^2}$. In this case, the second term of (5.17) is not zero and is equal to the mean residual vacation time $E[R_v]$. Thus, it follows from (5.17)

$$E[W] = \rho\,\{\overline{R} + E[W]\} + (1 - \rho)\,E[R_v]$$

Solving for $E[W]$ yields

$$E[W] = \frac{\rho\tau}{2\,(1 - \rho)}\left(1 + \frac{\delta^2}{\tau^2}\right) + \frac{\overline{V^2}}{2\,\overline{V}} \tag{5.18}$$

This result is an extension of (5.16) to the M/G/1 queue with server vacations.

5.5 The M/G/1 Queue with Priority Discipline

If the customers in a queueing system are put into different priority classes, the usual convention is to number the priority with an index, such that the smaller the index, the higher the priority; that is, customers of priority 1 are served before customers of priority $j \geq 2$.

Within each class, there are different service disciplines. The most common is head-of-the-line (HOL) service, which means that customers are served in order of arrival in a given class.

There are two other refinements possible in priority queueing:

1. Preemptive and
2. Non-preemptive

In preemptive queueing, a customer with higher priority is allowed to interrupt the service of a customer with lower priority. Such an interruption leads to a further classification based on whether the interrupted service is to be continued from the point of interruption at a later time or be restarted from the beginning. In non-preemptive queueing, there is no interruption. An arriving customer with higher priority must wait until the service in progress is completed.

(A) The HOL Non-Preemptive Priority System

We shall consider the HOL non-preemptive priority queueing system with m priority classes and derive an expression for the mean waiting time W_j of a test customer with priority index j, $1 \leq j \leq m$. Fig. 5-4 illustrates a non-preemptive HOL priority queueing system.

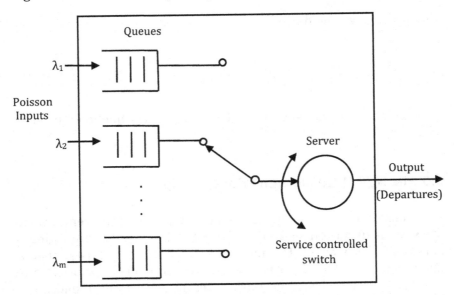

Fig. 5-4. Non-preemptive head-of-the-line (HOL) priority queueing system.

Each priority queue operates on a first-come, first-served (FIFO) queue discipline while being served. After a service is completed, the switch will move to the highest priority queue that has customers for service.

Suppose that customers arrive at each priority queue from independent Poisson processes with respective arrival rate λ_i, i = 1, 2, ..., m. Let $\overline{X}_i$ and $\overline{X}_i^2$ be the first and second moments of the service time X_i. Then, by definition, the offered traffic to priority queue i is

$$\rho_i = \lambda_i \overline{X}_i, \ i = 1, 2, \ldots, m$$

and under the equilibrium condition

$$\sum_{i=1}^{m} \rho_i < 1$$

It is important to note that the mean waiting time of a test customer of priority j consists of three components:

1. The average delay, if any, due to a customer already in service, W_0;

2. The average delay due to customers of higher or equal priority that arrive before the test customer and wait in the queue, W_A; and

3. The average delay due to customers of higher priorities that arrive after the test customer, W_B.

In what follows we shall calculate the mean value of each of these three components. First, by definition, W_0 is precisely the mean value of the residual service time $R_i = \overline{X}_i^2 / 2\overline{X}_i$. Since the probability of a priority i customer being served is ρ_i, the average delay W_0 is thus given by

$$W_0 = \sum_{i=1}^{m} \rho_i \; \frac{\overline{X}_i^2}{2\overline{X}_i} = \sum_{i=1}^{m} \frac{\lambda_i \overline{X}_i^2}{2}$$

Now consider the average delay W_A due to customers of higher or equal priority that are in the system before the arrival of the test customer of priority j. Let N_{ij} denote the number of such customers in the priority class i ≤ j. Since each of the customers has a mean service time $\overline{X}_i$, the total average delay W_A from all priority classes with index i ≤ j is simply

$$W_A = \sum_{i=1}^{j} \overline{N}_{ij} \; \overline{X}_i .$$

Using (3.23), we can write

$$\overline{N}_{ij} = \lambda_i W_i .$$

It follows that

$$W_A = \sum_{i=1}^{j} \rho_i W_i$$

Finally, we consider the delay caused by customers of priority i < j who arrive after the test customer. Let M_{ij} denote the number of such customers. Then, using (3.23) we can write

$$W_B = \sum_{i=1}^{j-1} \overline{M}_{ij} \; \overline{X}_i$$

$$= \sum_{i=1}^{j-1} \rho_i W_j$$

where $\overline{M}_{ij} = \lambda_i W_j$ and $\rho_i = \lambda_i \overline{X}_i$.

It follows that

$$W_j = W_0 + W_A + W_B$$

$$= W_0 + \sum_{i=1}^{j} \rho_i W_i + \sum_{i=1}^{j-1} \rho_i W_j$$

Solving this equation for W_j yields

$$W_j = \frac{W_0 + \sum_{i=1}^{j-1} \rho_i W_i}{1 - \sum_{i=1}^{j} \rho_i}, \quad j = 1, 2, \ldots, m$$

From this expression, we calculate

$$W_1 = \frac{W_0}{1 - \rho_1}$$

and

$$W_2 = \frac{W_0}{(1 - \rho_1 - \rho_2)(1 - \rho_1)}$$

Continuing this process, we find the following general relation:

$$W_j = \frac{W_0}{(1 - \delta_j)(1 - \delta_{j-1})}, \quad j = 1, 2, \ldots, m \tag{5.19}$$

where $\delta_j = \sum_{i=1}^{j} \rho_i$ with $\delta_0 = 0$.

This result is known as Cobham's formula.

(B) The Preemptive Priority System

In a preemptive priority system, an arriving customer with higher priority is allowed to interrupt the lower priority customer being served. The most natural way to handle the interrupted service is that the interrupted customer will resume his service from the point of interruption. This is called the preemptive resume strategy.

Since customers of the highest priority (class 1) are not affected by customers of lower priority classes, W_1 can be obtained immediately by applying the Pollaczek-Khinchin formula (5.16):

$$W_1 = \frac{\rho_1 \tau_1}{2(1 - \rho_1)} \left(\frac{\overline{X_1^2}}{\tau_1^2} \right)$$

$$= \frac{\lambda_1 \overline{X_1^2}}{2(1-\delta_1)}$$

where $\rho_1 = \lambda_1 \tau_1$ and $\overline{X_1^2}$ is the second moment of the service time of customer of class 1.

However, an arriving class j customer will have to wait first for the average delay W_{0j}^* due to a customer of class $i \le j$ already in service. Similar to the non-preemptive case, this quantity now becomes (see (5.19) for the difference)

$$W_{0j}^* = \sum_{i=1}^{j} \frac{\lambda_i \overline{X_i^2}}{2}, \quad j = 1, 2, \dots, m. \tag{5.20}$$

where the summation on the right-hand side only goes up to j.

Following exactly the same reasoning as the non-preemptive case to obtain (5.19), we write

$$W_j^* = W_{0j}^* + W_A^* + W_B^*$$

$$= W_{0j}^* + \sum_{i=1}^{j-1} \rho_i W_i^* + \sum_{i=1}^{j} \rho_i W_j^*$$

Solving the equation for W_j^* yields

$$W_j^* = \frac{W_{0j}^* + \sum_{i=1}^{j-1} \rho_i W_i^*}{1 - \sum_{i=1}^{j} \rho_i} \quad , \quad j = 1, 2, \ldots m$$

From this expression, we find

$$W_j^* = \frac{W_{0j}^*}{(1 - \delta_j)(1 - \delta_{j-1})}, \quad j = 1, 2, \ldots, m \qquad (5.21)$$

where $\delta_j = \sum_{i=1}^{j} \rho_i$, $\delta_0 = 0$, and W_{0j}^* is given by (5.20), which is the average waiting time of customers in the preemptive priority system on their first arrival. However, extra waiting time is introduced due to the fact that a customer of class j can be preempted (interrupted) by higher priority customers during service with probability δ_{j-1}, on average. Thus, the extra delay due to the service of a class j customer is $\delta_{j-1} \overline{X}_j$. Furthermore, this extra delay will introduce an additional delay equal to $\delta_{j-1} (\delta_{j-1} \overline{X}_j)$. Continuing this reasoning reveals that the extra average delay of a class j customer to complete his service is

$$(\delta_{j-1} + \delta^2_{j-1} + \ldots) \overline{X}_j = \frac{\delta_{j-1} \overline{X}_j}{1 - \delta_{j-1}}$$

Finally, the average waiting time for a class j customer is the sum of W_j^* in (5.21) and this extra delay:

$$W_j = \frac{\delta_{j-1} \overline{X_j}}{1 - \delta_{j-1}} + \frac{W_{0j}^*}{(1 - \delta_j)(1 - \delta_{j-1})} , \quad j = 1, 2, \dots, m \qquad (5.22)$$

where W_{0j}^* is given by (5.20).

5.6 The GI/M/1 Queue

This section will study the single server queue for which the interarrival times of customers are mutually independent and identically distributed, and the service time has an exponential distribution.

Suppose that customers arrive at time epochs $T_1, T_2, \dots$. We assume that the interarrival times

$$X_{k+1} = T_{k+1} - T_k , \quad k = 0, 1, \dots ; T_0 = 0$$

are mutually independent and identically distributed random variables and have the general distribution function

$$G(x) = P\{X_{k+1} \leq x\} , \quad k = 0, 1, \dots$$

and mean $1/\lambda$. The service time has an exponential distribution with mean $1/\mu$.

Let N_k denote the number of customers in the system just prior to the arrival of the kth customer, that is, N_k denotes the number of customers present at time $T_k - 0$. Fig. 5-5 depicts the (k+1)th interarrival time X_{k+1} and the Poisson departure process (see Section 1.7).

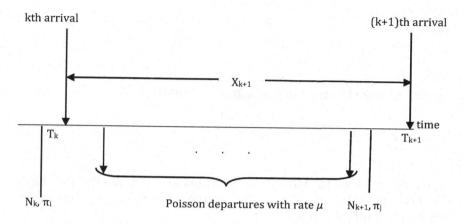

Fig. 5-5. The kth interarrival time and the departure process.

Applying the formula of total probability, we can write the state probability

$$P\{N_{k+1} = j\} = \sum_{i=0}^{\infty} P\{N_{k+1} = j \mid N_k = i\} \, P\{N_k = i\}$$

Since the conditional probability depends on the indices i and j but not on the index k, we let

$$p_{ij} = P\{N_{k+1} = j \mid N_k = i\}$$

Under the equilibrium condition $\rho = \lambda / \mu < 1$, an unique stationary state probability distribution

$$\pi_j = \lim_{k \to \infty} P\{N_k = j\}$$

exists. It follows from the formula of total probability that as $k \to \infty$, we have, in the steady state

$$\pi_j = \sum_{i=0}^{\infty} p_{ij} \, \pi_i \, , j = 0, 1, \ldots$$

which is subject to the normalization condition

$$\sum_{j=0}^{\infty} \pi_j = 1$$

We now proceed to determine the transition probability p_{ij}. First, observe that if $N_k = i$ and $N_{k+1} = j$, then the value of j cannot be greater than i+1. It follows that

$$p_{ij} = 0, j > i+1.$$

For $j \le i+1$, note that the server has been busy continuously during the interarrival time X_{k+1}, the departure process obeys the Poisson distribution with rate μ because the service times are exponentially distributed with mean $1/\mu$. It follows that the probability of exactly i−1+j departures during a given time interval of length x is

$$P\{N_{k+1} = j \mid N_k = i, X_{k+1} = x\} = \frac{(\mu x)^{i+1-j} \, e^{-\mu x}}{(i+1-j)!}$$

Since X_{k+1} has the distribution function $G(x)$, then the transition probability p_{ij} is given by

$$p_{ij} = \int_0^{\infty} \frac{(\mu x)^{i+1-j}}{(i+1-j)!} \, e^{-\mu x} \, dG(x) \, , i \ge j-1 \, , j = 0, 1, \ldots$$

It follows that

$$\pi_j = \sum_{i=j-1}^{\infty} \pi_i \int_0^{\infty} \frac{(\mu x)^{i+1-j}}{(i+1-j)!} \, e^{-\mu x} \, dG(x) \, , i \ge j-1$$

$$= \sum_{k=0}^{\infty} \pi_{k+j-1} \int_0^{\infty} \frac{(\mu x)^k}{k!} e^{-\mu x} dG(x) , \; j = 0, 1, ...$$

and the state probability distribution is now completely determined.

Recall that the state probability π_j is defined as the probability that the system is in state j right before the arrival epoch of customers. In fact, the arrival epochs here are regeneration points. To find π_j, we assume a solution of the form

$$\pi_j = A \, \delta^j , \; j = 0, 1, ... , \; 0 < \delta < 1$$

where A and δ are unknown quantities and remain to be determined. Using this solution in the last expression yields

$$\delta = \int_0^{\infty} e^{-(1-\delta)\mu x} dG(x) = \hat{G}((1-\delta)\mu) \qquad (5.23)$$

where $\hat{G}((1-\delta)\mu)$ is the Laplace –Stieljes transform of the function $G(x)$. The function $\hat{G}((1-\delta)\mu)$ has the following properties:

$$\hat{G}(\mu) = \hat{G}((1-\delta)\mu) \Big|_{\delta=0} = \int_0^{\infty} e^{-\mu x} dG(x) > 0$$

and

$$\hat{G}(0) = \hat{G}((1-\delta)\mu) \Big|_{\delta=1} = \hat{G}(0) = 1$$

Further, we also have

$$\frac{d\hat{G}}{d\delta} \Big|_{\delta=1} = \frac{\mu}{\lambda} = \frac{1}{\rho} > 1$$

Thus, an unique real solution δ_0 exists for δ in the range $0 < \delta < 1$. This is shown in Fig. 5-6.

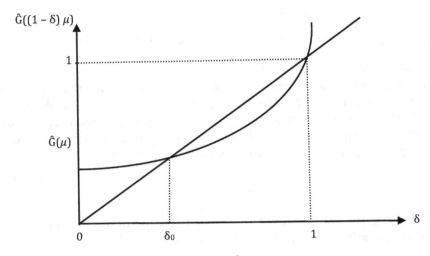

Fig. 5-6. Real solution of $\hat{G}((1 - \delta) \mu) = \delta$.

It remains to determine the constant A. Using the normalization condition gives

$$\sum_{j=0}^{\infty} \pi_j = \sum_{j=0}^{\infty} A \, \delta^j = 1$$

Hence,

$$A = 1 - \delta$$

and the state probability distribution is then given by

$$\pi_j = (1 - \delta) \, \delta^j , j = 0, 1, \ldots \tag{5.24}$$

where δ is given by the real solution δ_0 of (5.23).

(A) The Probability of Waiting and the Mean Waiting Time

The other quantities of practical interest are the probability of waiting and the mean waiting time. We let W denote the waiting time. If an arriving customer finds the server busy, he must wait. Thus, the probability of waiting is simply

$$p_w = P\{W > 0\} = 1 - \pi_0 = \delta \tag{5.25}$$

Using exactly the same reasoning as in the derivation of the Pollaczek-Khinchin formula for the mean waiting time, except that the probability of waiting is δ now instead of ρ and the $E[R]$ is simply the mean service time τ because of the exponential service time, we immediately obtain the mean waiting time for the GI/M/1 queue

$$E[W] = \frac{\delta\, E[R]}{1 - \delta} = \frac{\delta\, \tau}{1 - \delta} \tag{5.26}$$

(B) The Waiting Time Distribution Function

Following exactly the same reasoning as the investigation of the waiting time distribution function for the M/M/1 queue, we write

$$P\{W > t\} = \sum_{k=1}^{\infty} \pi_k\, P_k\{W > t\}$$

$$= \sum_{k=1}^{\infty} (1 - \delta)\, \delta^k \int_t^{\infty} \frac{\mu^k\, x^{k-1}}{(k-1)!}\, e^{-\mu x}\, dx$$

$$= \delta\, e^{-\mu(1-\delta)t} \tag{5.27}$$

where $P_0\{W > t\} = 0$ and δ is the unique solution δ_0 of the equation

$$\delta = \hat{G}((1 - \delta)\mu)$$

in the range $0 < \delta < 1$.

Comments. Similar to the M/G/1 queue case, for the GI/M/1 queue, if the interarrival time distribution is exponential with mean $1/\lambda$, then the GI/M/1 queue becomes the M/M/1 queue. All results obtained for the GI/M/1 queue should become those for the M/M/1 queue. Now let us assume that the interarrival time distribution function is given by

$$G(x) = 1 - e^{-\lambda x}$$

Then the Laplace-Stieljes transform of G(x) in (5.22) becomes

$$\hat{G}[(1 - \delta)\,\mu] = \int_0^\infty e^{-\mu(1-\delta)x}\ dG(x)$$

$$= \int_0^\infty e^{-\mu(1-\delta)x}\ \lambda\ e^{-\lambda x}\ dx$$

$$= \frac{\lambda}{\mu(1-\delta)+\lambda}$$

And (5.23) becomes

$$\delta = \frac{\lambda}{\mu(1-\delta)+\lambda}$$

Hence, we find

$$\delta = \frac{\lambda}{\mu} = \rho$$

which is the real solution of (5.23).

The state probability distribution in (5.24) is then given by

$$\pi_j = (1-\rho)\,\rho^j,\ j = 0, 1, \ldots$$

which is the same as p_k in (5.1) for the M/M/1 queue. Other results of (5.25) to (5.27) follow immediately.

Review

The simplest queueing system is the one with Poisson input and a single exponential server. The formulas for calculating the performance indices are very simple.

When the service time distribution is general, only the Pollaczek-Khinchin formula for the mean waiting time is available. We have presented a direct derivation of this formula without using the concept of imbedded Markov chain.

Furthermore, formulas for the mean waiting time are extended to the case of server vacations and the case of priority discipline.

For the system with a general input process and a single exponential server, formulas of the probability of waiting, the mean waiting time and the waiting time distribution function are obtained. These closed form formulas are expressed in terms of a real solution of a functional equation (5.23), which is related to the distribution function of the general input process. If the general input process is a Poisson input process, then the real solution will become the offered load to the system.

BIBLIOGRAPHY

[1]   Kendall, D.G., "Some Problems in the Theory of Queues", J. Roy,
       Statist. Soc. (B), 1951, Vol. 13, 151 – 185.

[2]   Feller, W., An Introduction to Probability Theory and its
       Applications, Vol. 1, 3rd edn., Wiley, 1968.

[3]   Khinchin, A.Y., Mathematical Methods in the Theory of
       Queueing, 2nd edn., Hafner Publishing Co., 1969.

[4]   Kleinrock, L., Queueing Systems: Vol. 1 - Theory, Wiley, 1976.

[5]   Cooper, R.B., Introduction to Queueing Theory, 2nd edn.,
       Elsevier Science, 1981.

[6]   Erlang, A.K., "Solution of Some Problems in the Theory of
       Probabilities of Significance in Automatic Telephone
       Exchanges", The Post Office Electrical Engineers J., 1917, Vol.
       10, 189 – 197.

[7]   Pinsky, E., Conway, A. and Liu, W., "Blocking Formulae for the
       Engset Model", IEEE Trans. Communications, 1994, Com – 42,
       2213 – 2214.

[8]   Wolft, R.W., Poisson Arrivals See Time Average, Opns. Res.,
       1982, Vol. 30, 223 - 231.

[9]   Chan, W.C. and Lin, Y.B., "Waiting Time Distribution for the
       M/M/m Queue", Proc. IEE – Communications, June 2003, Vol.
       150, 159 – 162.

[10]   Fry, T.C., Probability and Its Engineering Uses, 2nd edn., Van
       Norstrand, 1965.

[11] Chan, W.C., Performance Analysis of Telecommunications and Local Area Networks, Kluwer Academic Publishers, 2000.

[12] Khinchin, A.Y., Sequences of Chance Events with After-Effects, Theory of Probability and its Applications, 1956, Vol. 1, 1-15.

[13] Khinchin, A.Y., On Poisson Sequences of Chance Events, Theory of Probability and Applications, 1956, Vol. 1, 291-297.

[14] Beckmann, P., Elementary Queueing and Telephony, 1981, Rev. edn., Batavia, IL.: ABC Tele Training.

[15] Kleinrock, L., On the Modeling and Analysis of Computer Networks, 1993, Vol. 81, No.8, Proc. Of IEEE, 1179-1191.

[16] Chan, W.C., Modeling of Data Networks, International J. of Modeling and Simulation, 1993, Vol. 13, No.4, 121-128.

[17] Copenhagen Telephone Co., The Life and Works of A.K. Erlang, 1948.

[18] Redheffer, R.M., A Note on the Poisson Law, 1953, Vol. 26, No. 2, 185-188.

[19] Martine, R.R., Basic Traffic Analysis, Englewood Cliffs, N.J., Prentice Hall, 1994.

INDEX